Light-House Board United States.

Papers on the Comparative Merits of the Catoptric and Dioptric or Catadioptric Systems of Light-House illumination

And Other Subjects Relating to Aids to Navigation

Light-House Board United States.

Papers on the Comparative Merits of the Catoptric and Dioptric or Catadioptric Systems of Light-House illumination
And Other Subjects Relating to Aids to Navigation

ISBN/EAN: 9783337249496

Printed in Europe, USA, Canada, Australia, Japan

Cover: Foto ©berggeist007 / pixelio.de

More available books at **www.hansebooks.com**

PAPERS

ON THE

COMPARATIVE MERITS

OF THE

CATOPTRIC AND DIOPTRIC OR CATADIOPTRIC SYSTEMS

OF

LIGHT-HOUSE ILLUMINATION

AND OTHER SUBJECTS RELATING TO

AIDS TO NAVIGATION.

COMPILED FROM BRITISH, FRENCH, AND UNITED STATES REPORTS AND
AUTHORITIES, FOR THE USE OF THE UNITED STATES
LIGHT-HOUSE ESTABLISHMENT SERVICE.

WASHINGTON:
GOVERNMENT PRINTING OFFICE.
1861.

TABLE OF CONTENTS.

CATOPTRIC AND DIOPTRIC LIGHTS.

BRITISH PARLIAMENTARY REPORT, 1857.

CATOPTRIC AND DIOPTRIC LIGHTS.

PAPER BY MONS. LEONOR FRESNEL ON THE OLD AND NEW SYSTEMS OF
LIGHTS FOR LIGHT-HOUSES.

[REPORT of Lieut. THORNTON A. JENKINS and Lieut. RICHARD BACHE, U. S. Navy, "On
Improvements in the Light-house System and Collateral Aids to Navigation."—Senate
Doc., 1st session 29th Congress, No. 488, August, 1846, page 124.]

After having balanced the advantages relative to the two systems
of lights in view of their *useful* and *economical effects*, I ought to consider them with reference to their *security* and the *facility* with which
they are served.

I will reproduce, upon this subject, the observations inserted in a
memoir of the 20th of April, 1830, in which I replied to the questions which were addressed to me by the government of Sweden
and Norway, in relation to the necessary measures to be taken to
improve the lighting of its maritime coast :

"The service of lenticular lights is, in the aggregate, less laborious than that of the reflector lights. The first demand at all times
during the night the unremitted attention of the keeper. If, for
example, the central lamp should become extinct during the absence
of the keeper from the lantern, or while he is asleep, the horizon of
the light would remain some hours plunged in total darkness, and the
greatest objection which has been urged against our new system of
illumination is the fear of such accidents. Happily an experience of
seven years has dissipated that fear, and the lenticular lights have
been distinguished up to this time by the regularity of their service.
However, every precaution has, besides, been taken to replace
promptly the lamp or its burner in case of extinction. The extreme
simplicity of the day duty compensates the keepers for that to which

they are subjected during the night. To snuff and replace the wicks, renew the oil, sweep the chambers of the lantern and the stairs of the tower, dust the apparatus, and sometimes wash with a little spirits of wine the tarnished spots upon it, and, lastly, to wipe dry the glass of the lantern ; such is the principal daily duty which is divided between the keepers of the new lights, and which rarely occupies them more than two hours.''

Opinions thus expressed *fifteen years since*, based upon an experience of seven years, have been greatly strengthened up to the present time, embracing a period of *twenty-two years* since the establishment of the Cordouan light, and sustained by the results daily offered of more than *one hundred* lenticular lights of the three first orders, established along the coasts as well of France as of different foreign powers.

In this important point of view, then, the question seems to be irrevocably settled, and I will only add a few considerations relative to the application, more or less extended, which may be made of the new system of illumination to the vast maritime coasts of the United States.

First. It has been objected that it would require too great sacrifices to be made to procure in that country keepers possessing the amount of intelligence requisite for the superintendence of lenticular lights.

Second. That from distant points or stations the necessary repairing and renewing of the mechanical lamps would be attended with great difficulty.

I will reply, with regard to the *keepers*, that the difficulty of obtaining proper persons to fill these subaltern stations appears to be most singularly exaggerated. In France they belong almost always to the class of *ordinary mechanics*, or laborers, who make from 1.50 francs to 2.50 francs (27 to 46 cents) per day.

Eight or ten days will suffice, ordinarily, to instruct a light-keeper in the most essential parts of his duty, receiving lessons from an instructor conversant with all the details of service; and two instructing officers will be sufficient to prepare keepers for all the lenticular lights which could be successively established upon the coasts of North America. The information thus imparted would never be lost, and these officers might, besides, be aided by foremen or assistants, who could supply the place in case of necessity. In defence of this assertion I will cite the example of the administration of Norway and Sweden, which, after having obtained the assistance of a French agent to put up the apparatus of the two first lenticular lights, which were sent from Paris in 1832 and 1836, has provided since, without any foreign assistance, for the placing as well as the organization of the service, of all the lights of the new system which it has successively established.

With reference to the eventual repairs of the mechanical lamps, it is to be considered—

FIRST. That, in consequence of the great strength of the pieces of which the new model of mechanical lamps is composed, they will perform well for a number of years without requiring anything more than a proper attention to their cleanliness.

SECOND. That the ordinary assortment of a dioptric light-house comprises three of these lamps, which afford a sufficient guarantee against the chances of accident; and, besides, we may, by increasing a little the mean expense, increase the number to *four* under some circumstances, as an exception to the general rule.

THIRD. That the repairs of the implements under discussion may be easily made by all the clock or watchmakers, or other mechanicians, to whom we have recourse for repairing the revolving machinery of light-houses.

I conclude with the remark that, if it be determined to multiply the application of the new system of maritime illumination in the United States, it seems to me that it will be expedient to *engage one of the foremen employed in the manufactory of our mechanical lamps to go to, and remain in the country for several years.* By that measure, which would be attended with very little expense, all the difficulties which might presnet themselves at first in establishing lenticular lights would be removed, and the perfect regularity of the service of these new establishments would be insured.

LEONOR FRESNEL.

PARIS, *December* 31, 1845.

PAPER ON FRENCH LIGHTS ACCOMPANYING REPORT OF THE LIGHT-HOUSE
BOARD OF THE UNITED STATES.

[REPORT OF Lieuts. THORNTON A. JENKINS and RICHARD BACHE, U. S. Navy, to the Secretary
of the Treasury, June 22, 1846.—Senate Doc., 1st session 29th Congress, No. 488, p. 70.]

The Light-house Department of France is attached to the official
duties of the minister, Secretary of State for the Interior, and is
under the immediate control and direction of the Minister of Public
Works, charged with the administration of the bridges and roads.

A central public board has the management of all light-houses,
buoys, beacons, and sea-marks on the coasts, which is composed of
eleven distinguished scientific and professional individuals, who are
appointed by the government, including the engineer, secretary to
the commission, and his assistant. This board is presided over by
the Minister of Public Works, and in his absence by the Under Secretary of State for that department.

This mixed commission, called the "Commission des Phares," is
composed of naval officers, (of whom there is a majority,) of inspectors
of the corps of bridges and roads, and of members of the institute.
It prepares the projéts for all new lights, and the general council of
bridges and roads judges of the propriety of all schemes for that
branch of service, under the four heads of Architectural Design,
Mode of executing the Works, Estimate of the Expense, and the
Preparation of the Specifications of the Works. The light-house
commission of France is not an administrative body, but is occupied
solely in questions of principle or design, and leaves to the general
directory of bridges and roads the care of providing the necessary
means for the construction of new works, the expenses of illumination, &c.

The central commission at Paris is charged with the duty of providing all supplies necessary for keeping the illuminating apparatus
in perfect order. There is also in Paris, belonging to this particular
branch of the public service, a central workshop and depot, under
the immediate care and supervision of the secretary-engineer to the
commission, who superintends the construction (by mechanics employed by the administration) of all lanterns and their fixtures that

may be required for the service ; tests all apparatus before sending
it to its destination ; makes experiments upon all the optical and
mechanical portions of apparatus destined for light-house purposes,
combustibles, &c.; in short, this officer is charged with all the scientific details of the service, subject to the instructions, from time to
time, which may be issued by the light-house commission. At this
central depot are always kept, ready for immediate use, the various
articles required in the illuminating department ; such, for example,
as mechanical and Argand lamps, glass chimneys, wicks, cleaning
materials, &c.; also specimens of the different descriptions of apparatus used in light-houses, and apparatus constructed upon the latest
and most approved plans ready for service.

All expenses incurred in the maintenance of the lights and their
appendages are defrayed by the agents of the national treasury,
from funds authorized by annual appropriations for those specific
purposes.

No light dues are charged upon shipping in France, as in Great
Britain, Holland, Denmark, Norway, and Sweden, &c. ; but the
whole establishment is provided for as in the United States and
Russia.

The maintenance of the light-house buildings is confided to the
departmental or local engineers, and the expenses are defrayed from
funds appropriated for the service of the department of public works.

The establishment of new works is decided upon by the Minister
of Public Works, under the advice of the light-house commission.
The determination of the minister is reported officially by the secretary of the commission to the Under Secretary of State for that
department, and through his office to the Prefect of the department
in which the proposed work is to be established. The Prefect directs
the chief engineer of bridges and roads for that department to have
detailed plans and estimates prepared upon the basis of the proposition of the light-house commission ; these plans and estimates are
transmitted through the office of the Under Secretary of State to the
secretary of the light-house commission, who makes a report to
accompany them to the light-house commission. The plans and
estimates are then submitted to the light-house commission, which
decides whether or not the wants of the service, *nautically* or *otherwise*, are such as to require the construction of the proposed works.
In the preparation of these plans and estimates, the military engineer
of the department is consulted, to ascertain his opinions as to the

propriety of constructing these works with reference to the defences of the coast.

The details having been completed, after having undergone the strictest scrutiny in every particular, the projét is presented to the general council of bridges and roads, to be considered with reference to the architectural designs, mode of construction, estimate of expense, &c. Having been approved by the general council of bridges and roads, and the Minister of the Interior, the plan is then sent to the Prefect of the department in which the light is to be established, with instructions to enter into contracts for the execution of the works, under the specifications and limitations authorized by the administration.

The execution of these works is entrusted to the engineers of bridges and roads for that department. As the works advance, the contractor receives payments upon the certificates of the engineers in charge, approved by the Prefect of the department, from the departmental paymaster, (as deputy of the public treasury,) and the sums are charged to the budget for works of navigation, under the head of light-houses.

The light-house towers of France are constructed in the most substantial and perfect manner possible, without there being any appearance of unnecessary or wasteful expenditure. Great care is taken in the interior arrangements of the buildings, so that they may best answer the requirements of the service. Many of the towers are constructed of a soft stone of a rather peculiar kind, which hardens by exposure to the action of the atmosphere : those constructed of that material are lined inside with brick, leaving a sufficient space between the interior of the outer wall and the brick to allow a free circulation of air, thereby securing the building from dampness. Hard burnt bricks are preferred for light-house towers, when circumstances will admit of their being employed, particularly in fitting up the oil apartments, which are placed below the surface of the earth, to insure as equable a temperature during the whole year as may be possible to attain. The keepers' apartments are finished and fitted up in a plain, substantial and economical manner, combining all the necessary accommodation and comfort. There is a room fitted and properly furnished for the accommodation of the engineer, inspector, or other person authorized to make official visits, at each light-station. Especial care is taken to secure proper ventilation to the towers and lanterns—all the necessary fixtures about the light-rooms, lanterns, apparatus, &c.—the most minute, and apparently unimportant details in the exterior and interior arrangements ; in short, nothing could combine greater perfection in stability, in usefulness, and a proper economy, than is perceptible in everything connected with the light-houses visited by us on the coasts of France.

The repairs of the light-houses and their appendages are projected and executed by the engineers of the different departments in which they exist, who are limited as much as possible in their expenditures by the estimates of each year for those specific purposes. In some cases the contractor general is authorized to make repairs, under the direction of the agents of the administration of bridges and roads.

Whenever application is made for a new harbor light, the subject is submitted to a local commission, assisted by the engineers of the department. The report is discussed by the light-house commission, and the same course subsequently followed as in the case of large or seacoast lights.

All the light-house towers in France are furnished with lightning conductors, made of copper wire twisted in the form of a rope, and about three-fourths of an inch in diameter.

In the organization of the lighting service, two systems are followed : the contract, and the administrative. The Ocean and Mediterannean coasts are under contract at present for nine years from 1839, for all the detail supplies of the service. 　*　*　*　*　

*　　*　　Among the clauses and conditions, it will be perceived that the contractor general is required to be represented by a deputy in each department in which there are any lights ; that the *oil of colza, clarified and refined*, must be used exclusively ; and that the prices of oil will be regulated quarterly, based upon the average prices of the principal market in the kingdom for that particular article of commerce. M. Fresnel insists that this last clause has had a most salutary effect of insuring the best oil the market could produce, without the contractor running any risk of loss. On the coast of the channel, from the frontier of Belgium to St. Malo, this service is performed by the administration, except for the article of oil, which is procured under a contract entered into for three years. That portion of the coasts of France which is lighted by contract includes even the salaries of the light-keepers ; but where the service is performed by the administration, the keepers are appointed by the Prefect of the department, upon the recommendation of the engineers. The smaller articles necessary to the illumination are

sent from the central depot in Paris, under the charge of a conducting steward. The mechanical lamps are sent to Paris to be repaired under the engineer secretary to the light-house commission. The administrative system recommends itself, for the reason that it avoids all intervention of interest foreign to that for which the lights were established. The contract system has been for a long time preferred in France, for reasons of economy, complication of accounts when performed by the administration, &c.; but the experience of the last seven years on the channel coast has sufficiently demonstrated the importance of changing it to the administrative ; and it is deemed quite probable, that, after the expiration of the present leases, that system will be exclusively adopted, except for supplies of oil.

The superintendence of the lights of France is confided to the local engineers of the corps of bridges and roads. The secretary to the light-house commission visits, each year, one of the three divisions into which the coast is divided, and his assistant another, so that the inspections, as far as possible, are biennial for each division. Monthly returns are made of all stores on hand, of the quantity of oil consumed each night, &c., to the secretary of the commission. These returns are intended as checks upon the keepers and answer the purpose admirably. A most rigid supervision is required at the hands of the inspecting engineers ; and, moreover, that they employ all possible means to detect any delinquency on the part of the keepers, or other agents connected with the service. It is conceded that all these precautions may fail to produce the desired effect, but that under such a supervision few among the guilty will escape detection. The lights visited by the undersigned were clean, and presented every indication of a perfect and systematic attendance and supervision.

Indications of the range of visibility afford very meagre data for forming a correct idea as to the relative value of apparatus for illumination. It is impossible to determine with certainty the *absolute range* of any light, in consequence of the different conditions of the atmosphere, and of the capacities of the different observers. A first order dioptric light has been seen *fifty* miles very often, and one of the fourth order as far as sixteen miles. M. Fresnel says upon the subject of range : "We would, then, draw very erroneous conclusions as to the relative value of the useful effect of the apparatus of these lights, in taking for a basis of comparison the indications of *range*, which are never fixed or positive."

At the present time there are two systems of illumination in France —the old or reflector system, and the new or dioptric system. In 1822 M. A. Fresnel placed the first dioptric apparatus ever successfully employed in the tower of Cordouan, at the mouth of the Gironde. In 1825 the light-house commission decided upon the exclusive use of the lenticular apparatus for the illuminations of the coasts of France and colonies; adopting, at the same time, the programme and report of Rear Admiral de Rossel, who had been charged, as a member of the "Commission des Phares," with that service.

Since that period new lights have been established, and old ones replaced with this new apparatus, until, on the 31st December, 1845, there were, of the two hundred and nine lights of every description belonging to the light-house department of France, one hundred and nineteen fitted with that apparatus. The remaining ninety lights were reflector lights, fitted with the Bordier Marcet (called "sideral") reflectors, and the parabolic reflectors, similar to those used in Great Britain and America. Of these last ninety lights, seventy-seven are small harbor or temporary lights, fitted in most cases with a single parabolic or Bordier Marcet reflector, marking the entrance to some channel or harbor. The remaining thirteen are fitted with lenticular apparatus of the most approved construction, in accordance with the original plan of 1825.

Engineers and other scientific and philanthropic individuals, of most, if not of all, the nations of the world, have made this new system of illumination an object of study and of critical examination; the results of which have been the successful, though gradual, application of it to the coasts of nearly all the commercial nations.

On the 31st December, 1845, *eighty-three* light-houses belonging to foreign governments had been fitted with lenticular apparatus, constructed in Paris; to which may be added those constructed in England and Holland, say from fifteen to twenty, making, including those on the coasts of France, upwards of two hundred and ten; one hundred of which may be put down as of the three first orders, and the remaining one hundred and ten of the fourth order. These numbers do not include those at present in the course of construction for France, Egypt, (tower at Alexandria,) Brazil, and the colonies, islands in the Pacific, &c. M. Fresnel says, with perfect truth and reason, "After these numerous and extended applications, the dioptric system of lights may be fully appreciated under the double aspect of *theory* and *practice:* and I will add, that under the first point of view the question has been for a long time out of controversy."

There are six different orders of lenticular apparatus at present employed, viz : first, second, third larger model, third smaller model, fourth larger model, and fourth smaller model order.

The different orders are subjected to different combinations, such as dioptric, two catadioptric, one with concave mirrors, and the other with catadioptric zones, or rings of glass, in triangular profile sections, and the "diacatoptric," * combining the dioptric portion and the catadioptric zones surmounted by plane mirrors. In addition, a spherically curved metallic reflector or mirror is placed on the land side of all lights which are only required to illuminate from four-fifths to five-sixths of the horizon, which reflects the rays from that side back through the opposite lenses.

"There can be no doubt," says a distinguished engineer,† who has had much to do with the light-houses of Europe, "that the more fully the system of Fresnel is understood, the more certainly will it take the place of all other systems of illumination for light-houses, at least in those countries where this important branch of administration is conducted with the care and solicitude which it deserves."

"To the Dutch belongs the honor of having first employed the system of Fresnel in their lights." "The commissioners of northern lights followed in the train of improvements, and in 1834 sent Mr. Alan Stevenson on a mission to Paris, with full powers to take such steps for acquiring a perfect knowledge of the dioptric system, and for forming an opinion of its merits, as he should find necessary."

"The singular liberality with which he was received by M. Leonor Fresnel, brother to the late illustrious inventor of the system, and his successor as secretary to the light-house commission of France, afforded Mr. A. Stevenson the means of acquiring such information and making such a report, on his return, as to induce the commissioners of northern lights to authorize him to remove the reflecting apparatus of the revolving light at Inchkeith, and substitute dioptric instruments in its place." *   *   *   *

"The Trinity House followed next in adopting the improved system." *   *   *   *   Other countries begin to show symptoms of interest in this important change ; and America, it is believed, is likely soon to adopt active measures for the im-

* See Mr. Alan Stevenson's Report to Commissioners of Northern Lights for this word.
† Mr. Alan Stevenson, civil engineer.

provement of their light-houses. "Fresnel, who is already classed with the greatest of those inventive minds which extend the boundaries of human knowledge, will thus, at the same time, receive a place amongst those benefactors of the species who have consecrated their genius to the common good of mankind ; and wherever maritime intercourse prevails, the solid advantages which his labors have procured will be felt and acknowledged."

The fourth order lenticular lights are illuminated ordinarily by means of a common fountain, or constant level lamp and Argand burner, with a single cylindrical wick of three-fourths to seven-eighths of an inch in diameter, consuming about one and a quarter ounce of oil per hour, and forty-eight gallons per annum. The larger lights require mechanical lamps with multiple wicks, to as great a number as four, placed in concentric tubes, and the oil supplied to them by means of pumps, put in play by clock machinery. Hydraulic and pneumatic lamps have been employed in the place of the mechanical ones, but, with good reason, they are not approved of in France. For the catrdioptric apparatus of half a metre in diameter, the ordinary constant level lamps, with two concentric wicks, burning about four and a half ounces of oil per hour, have been employed very successfully at several points on the coast of France, where the ordinary range of a light of the third order, for example, was not required, or for harbor lights requiring a powerful ray, or one whose brilliancy it is necessary to weaken by the application of a red chimney, with the view to give it a distinctive character. These double wick ordinary lamps require only one keeper to attend to them. Some of the burners in France are fitted with flat wicks for small and temporary lights, although by no means common, and generally disapproved of.

The dioptric lights of France are divided into six different orders ; but, with reference to their distinctive characteristics and appearances, this division does not apply, inasmuch as, in every order or class, lights of precisely the same character may be found, differing only in the distance at which they can be seen, and in the expense of their maintenance. The six different orders, as before mentioned, are not intended as distinctions, "but are characteristic of the power and range of lights, which render them suitable for different localities on the coasts, according to the distance at which they can be seen."

"This division, therefore, is analogous to that which separates the lights of Great Britain into sea lights, secondary lights, and harbor lights, terms which are used to designate the power and position, and not the appearance of the lights to which they are applied."

In France there are nine principal combinations of lights possessing distinctive characteristics. These distinctions, for the most part, depend upon the periods of revolution rather than upon the characteristic appearance of the light. They are—

1. Flashes, which succeed each other every minute.
2. Flashes, which succeed each other every half-minute.
3. Flashes, alternately red and white.
4. Fixed lights, varied by flashes every four minutes.
5. Fixed lights, varied by flashes every three minutes.
6. Fixed lights, varied by flashes every two minutes.
7. Fixed white lights, varied by red flashes more or less frequent.
8. Fixed lights.
9. Double fixed lights.

There are very few double fixed lights in France. They are, however, sometimes employed for the purpose of giving a very decided character to the locality. For example, the first order lights at La Hève, near the port of Havre, and the two lights at present in the course of construction on the left bank of the *Canche*. Red fixed lights are not employed on the coasts of France, except as a distinguishing characteristic for harbor purposes. They are doubly objectionable : first, because of the great diminution of light in consequence of the absorption of the red glass chimney ; and, secondly, it loses its distinctive character in foggy weather, all lights assuming a reddish tint under those circumstances.

The revolving reflector lights are objected to because of the fact that, ordinarily, they are only distinguishable by the duration of their eclipses, which often become positive at a very short distance from the light-house, and the interval of time between any two eclipses could not be extended to a greater limit that three minutes without prolonging the duration of the eclipses to such an-extent of time as to mislead the navigator by depriving him for so long a time of his point of recognition. In the revolving dioptric apparatus, upon the latest and most approved plan, the duration of the eclipses is scarcely perceptible, the fixed subsidiary parts of which reflect a light constantly visible in a horizon extending nine or ten nautical miles with a second order, and from twelve to fifteen with a first order apparatus.

The three first of the principal combinations only are applied to the first three orders, in consideration that in the inferior orders the flashes would have too short a duration, and the eclipses would be positive at too short a distance from the light, in consequence of the feebleness of the ray produced by the fixed subsidiary part of the apparatus.

The distinguished engineer, secretary to the "Commission des Phares" of France, M. Leonor Fresnel, kindly furnished the undersigned with the results of numerous photometric experiments which were made for the purpose of testing the comparative useful and economical effects of the two systems of illumination, to which they beg leave to call particular attention.

M. Fresnel says, in his note referred to, "the foregoing results confirm the following principles :

"1. The *useful effect* of a parabolic reflector increases with its dimensions, and with that of the illuminating body.

"2. The *economical effect* of a reflector of given dimensions is greatest when the lamp-burner is smallest.

"The *divergence* is greatest when the flame is most voluminous, or when the reflector is smallest. We cannot then (all other things being equal) augment the *economical effect* of a reflector without diminishing its useful effect ; that is to say, without reducing its brilliancy or intensity, and consequently its range (portée).

"The reduction of the volume of light within certain limits is particularly objectionable when it appertains to eclipse apparatus, in which case it limits the width of the luminous cone, and consequently augments the length of the eclipses. The same reduction applied to the foci of reflectors composing a fixed light apparatus may weaken the light in their intervals to such a degree as to produce *dead angles*, or become completely obscured to the observer beyond certain distances.

"It is further proper to remark that the horizontal divergence is not lost for *useful effect*, but that the divergence in the vertical sense only profits the navigator in the limited angular space comprised between the tangent at the surface of the sea and the ray terminating at the distance of some miles from the light.

"Finally, there is for the calibre of the lamp burners applicable to reflectors of given dimensions, and destined for the illumination of an equally determined range, a *maximum* beyond which prodigality of light ensues, and a *minimum* within which the illumination becomes insufficient."

The third order smaller size lenticular apparatus may be illuminated with very decided advantages by means of an ordinary Argand burner and single wick. Such a light would consume about two ounces of oil per hour, and is admirably adapted for harbor lights. In ordinary weather such a light may be seen from twelve to fifteen miles. One keeper alone can attend to all the duties of such a light, and it is maintained in France at an annual expense of about two hundred dollars.

M. Fresnel remarks, with reference to the ranges of different lights, their useful effect, &c. :

"The useful effect of a light-house apparatus is measured by the quantity of light which it projects upon the horizon. Observations of *range* for that purpose furnish very uncertain evidences, on account of the difficulty of ascertaining the *absolute range* of a light, which varies according to the state of the atmosphere and according to the good or bad sight of the observers."

Reflector lights, with not more than six or eight burners, are attended by one keeper, occasionally assisted by the members of his family. For lights with a larger number of burners, two keepers ; and if the light be in an isolated position, three keepers are allowed, with, in the latter cases, certain privileges not accorded to others.

Dioptric lights of the fourth order and third order smaller size, require but one keeper, except when in isolated positions. Two keepers are allowed to lights of the third order larger size, and for those of the second order, in consequence of the employment of the mechanical lamp.

First order lights are allowed three keepers ; and when there are two first order lights forming one combination, five keepers are allowed for the two lights. Lights of the first order in isolated positions, are allowed four keepers, and for the third order larger size and the second order lights, similarily situated, three keepers are allowed.*

In comparing the two systems of illumination, they should be considered under the heads—first, of absolute, useful, and economical effect ; second, of first cost, repairs, and maintenance ; and, third, of the facility and safety of the service.

The brilliancy of a catadioptric apparatus of 11·8 inches in diameter, lighted by a lamp burning forty-five grammes of oil per hour,

---

° In England, Scotland, and Ireland, no difference is made between the number of keepers for dioptric and reflector lights.

has been found, by photometric experiments, to be equal to eight or nine Carcel burners; while that of a "sideral" reflector of Bordier Marcet, illuminated by a lamp consuming fifty grammes of oil per hour, has been found, in the same manner, equal to only four burners of Carcel; or in other words, the brilliancy of the former is to the latter as one to two. The useful effect of the catadioptric apparatus, illuminating three-fourths of the horizon, is represented by 137,700, and that of the reflector by 68,400, which gives the value as one to two.

The economical effect of the catadioptric apparatus is represented by 3060. and that of the reflector by 1368; giving the value in that respect as 1 to 2·24.

No combination of reflectors can produce an equivalent to the third order smaller size apparatus, illuminated by an ordinary fountain lamp and Argand burner, with one wick, consuming sixty grammes of oil per hour, or one burner, with two wicks, consuming one hundred and fifteen grammes of oil per hour. An apparatus of this sort, with a lamp of two wicks, may be seen in ordinary weather (the horizon of the light, from its elevation above the sea level, being equal to or greater than that distance) at the distance of fifteen to eighteen nautical miles.

The brilliancy of a catadioptric third order larger size apparatus, illuminated by a mechanical lamp of two wicks, consuming one hundred and ninety grammes of oil per hour (six and three-fourth ounces,) has been found equal to seventy burners.

We suppose that it embraces only four-fifths of the horizon. To illuminate, by means of reflectors, the same angular space of 288°, with an *effect of light about equal,* fourteen parabolic reflectors, of about eleven inches in diameter, illuminated by Argand lamps, consuming each thirty-five grammes of oil per hour, will be required. The useful effect of these reflectors will be represented by 870,240, and that of the catadioptric apparatus by 1,160,000; and thus it is seen, that notwithstanding the very great difference in favor of the catadioptric apparatus, in the consumption of oil, it is also superior in useful effect to the light with the fourteen parabolic reflectors. Further, the economical effect of the catadioptric apparatus is represented by 6105, and that of the reflector by 1776, or as 1 to 3·44: "that is to say, without estimating the expenditure of oil by *unity of light,* the lenticular light will be nearly three and half times more advantageous than the reflector light." With regard to the effective

2

expenditure of oil, they will be in the proportion of 190 grammes to 14 × 35 grammes per hour, or as 1 to 2.6.

The brilliancy of a catadioptric apparatus of the second order, with a mechanical lamp of three concentric wicks, consuming 500 grammes of oil per hour, has been found equal to 264 burners. Supposing that it is only required to illuminate three-fourths of the horizon, then, to obtain an effect about equal in angular space of 270°, at least 34 parabolic reflectors of about 20 inches in diameter will be required, which will give a useful effect which is represented by 3,525,120, while that of the catadioptric apparatus is represented by 4,120,000. The comparison between the absolute consumption of oil will be equal to 2·86 to 1, and that of the quantity of oil expended by unity of light equal to 3·33 to 1; thus, under this last report, the lenticular apparatus will be three and a third times as advantageous as the catoptric apparatus.

The maximum brilliancy of a revolving light of the second order, with 12 lenses, has been found to be equal to 1,184 burners, and its minimum brilliancy equal to 104 burners. To construct a light, with parabolic reflectors, possessing an equal effect, it will require 24 with diameters from 22 inches to 24 inches, arranged on six faces of the revolving frame. In making the comparison, however, for want of precise data as to the lustres of those reflectors, those of about 20 inches diameter will be referred to. It is supposed that the two lights compared are constructed so as to present the same distinguishing features; the maximum lustre of the reflector light will be equal only to 1,080 burners, with other disadvantages, for the details of which reference may be made to M. Fresnel's note No. 1, section 2, (hereto annexed.) M. Fresnel remarks, in this connection, "without pressing further the comparison of the effects of the two kinds of apparatus, we will perceive, without doubt, the evident advantages of the dioptric or lenticular combination, which in fine weather will not present *an absolute eclipse* at a less distance than from 15 to 18 nautical miles. If we now consider the expenditures of oil, we will find, first, that they are as 24 × 42 is to 500, or as 1 to 2 ; second, that the economical effects will be as 2,469 is to 10,043, or as 1 to 4·07; thus the lenticular apparatus will be four times as advantageous as the reflector apparatus.'' Let us remark, before proceeding further, that in employing 24 parabolic reflectors of about 20 inches diameter for such an apparatus we reach the utmost possible limit, without admitting the employment of lanterns of a size beyond all proper bounds; and we may also affirm that very few of the catoptric lights, considered as lights of the *first order*, equal the lenticular lights of the same character of the second order.

With reference to the first order dioptric lights, M. Fresnel remarks, in his note: "Now we have found that the total lustre or brilliancy of an apparatus of this kind is equal in all its azimuths to 480 burners of Carcel. But it will be practically impossible to obtain a like effect in the catoptric system, without having recourse to the employment of 36 parabolic reflectors of about 24 inches in diameter." "The difficulty becomes still greater, if it be necessary to attain with these reflectors the effect of a revolving lenticular light, with eight large lenses, the lustres or flashes of which exceed 4000 burners of the Carcel lamp."

"Let us limit ourselves, then, without entering into more full details, to the observation, that the *economical effect* of a fixed light of the first order, illuminating three-fourths of the horizon, is to the *economical effect* of a light composed of parabolic reflectors of about twenty inches diameter, as 10,080 to 2,469 or as 4·08 to 1: that is to say, that the first will be (as to the expense of the oil only) four times as advantageous as the second."

With regard to lights *varied by flashes* or *short eclipse lights*, "the catoptric system is not susceptible of producing that combination without great difficulty, which unites to the permanence of fixed lights the advantage of presenting a very decided character." No repairs are required upon the lenticular apparatus. * * * * *

The amount necessary to construct and put into operation "sideral" light for harbor purposes may be stated at 8,150 francs, or about $1,500; and the annual expense for its maintenance, including interest upon the cost at the rate of five per cent., at 1,207 francs, or about $225.

The amount necessary for a catadioptric smaller model harbor light may be put down at 9,181 francs, or about $1,700; and the annual expense for maintenance, including interest of first cost, &c., as above, at 1,259 francs, or about $235.

The useful effect of the "sideral" light has been found equal to 68,400, and its economical effect represented by 57.

The useful effect of the catadioptric light, illuminating three-fourths of the horizon, has been found equal to 137,700, and its economical effect, after the same manner, is represented by 109. The comparison of these two will, then, be in the proportion of 57 to 109, or as 1 to 1·91. "Then, besides the advantages of *a double lustre*, the catadioptric apparatus, in an economical point of view, is nearly twice as advantageous as the catoptric apparatus."

M. Fresnel remarks: "It is difficult to establish a comparison of a precise kind between the fixed lights of the third order in the old and the new systems, because we cannot obtain with the ordinary parabolic reflectors a passably equal distribution of light, without multiplying those reflectors to such a number as would require a much greater expenditure of oil than could be allowed for lights of that class." He says further: "I will merely observe that I have every reason to believe, from the indications contained in the table of light-houses of the United States, that among all the lights of that country illuminated by reflectors, the diameters of which do not exceed sixteen English inches, "there are very few whose *useful effect* is superior or equal to that of a catadioptric light of the third order larger model."

The amount necessary for establishing a reflecting revolving light with twenty-four parabolic reflectors of about twenty inches diameter is estimated at 73,000 francs, or about $13,700.

Annual expense for maintenance of the same, including interest at five per cent. per annum, will be 8,650 francs, or about $1,625.

The amount necessary for establishing a second order revolving lenticular light is estimated at 105,500 francs, or about $19,800.

The annual expense for maintenance of the same, including interest at five per cent. per annum, will be 11,075 francs, or about $2,075.

The useful effect of the reflector light is represented by 2,488,320, and its economical effect by 288.

The useful effect of the lenticular light is represented by 5,021,467, and its economical effect by 453.

The economical effect of these two lights will then be represented by 288 and 453, or in the proportion of 1 to 1·6. "From whence it results definitively that the lenticular light of the second order will be more than *one and a half times* as advantageous as the catoptric or reflector light, which we may without doubt consider as being of the first order, and the *useful effect* of which, nevertheless, could not be equal to *but half* of the *useful effect* of the former."

No comparison can be entered into between the first order lenticular lights and reflector lights, for the reason that it is impossible to construct a reflector light which would produce a sufficiently powerful effect to be compared to a dioptric one, without increasing the dimensions of the lantern, and the number and size of the reflectors, to a degree which would be attended with a very great expense, and equally great inconvenience.

From the foregoing details, which have been drawn mainly from information furnished by M. Fresnel, the following seems to be but just conclusions:

"1. That the lights fitted with the dicptric apparatus present a variety in their power and effects, and may be made to produce an intensity of lustre, which render them of an interest, in a nautical point of view, incontestably superior to those fitted with the catoptric apparatus.

"2. That if we take into account the first cost of construction and the expense of their maintenance, we will find, with respect to the effect produced, the new system (dioptric) is still from *once and a half* to *twice* as advantageous as the old (reflector.)"

If additional arguments and evidence were wanting to establish the now almost universally conceded fact, of the very positive and decided advantages of the dioptric system of Fresnel over all other modes of illumination for light-houses, they might be found to exist at present in an unanswerable form—that of the practical and successful application of the system, within the last few years, in nearly all the commercial nations of the world. Prior to the year 1832, there was not a single dioptric light out of France; and on the French coast, at as late a period as 1834, there were but 14 large and 15 small, or harbor lights, fitted with the dioptric apparatus.

On the 31st December, 1845, there were belonging to the French light-house department *one hundred and twelve lights* fitted with the dioptric apparatus, and throughout the world not less than *two hundred and ten lights* fitted upon this new system ; one hundred of which are of the three first orders, and the remaining one hundred and ten, small or harbor lights, without including apparatus now in course of construction at Paris, to which allusion has already been made.

The objections which have been made by *a few persons* to the employment of the Fresnel dioptric apparatus for the illumination of light-houses, in consequence, as they allege, of the difficulties which attend the management of the mechanical lamps with concentric wicks (which are absolutely necessary for the proper illumination of the larger orders of apparatus) seem to be no longer tenable, if indeed there ever were any reasonable grounds of objection on that account.

The twenty-three years' experience in France (dating from the time the Cordouan light was exhibited,) where *ordinary day laborers are taken for light-keepers*, and the undeniable fact of the successful employment of the system for fourteen years in Holland, Scotland, and Norway; for from five to ten years in England, Sweden, Denmark, Prussia, Belgium, Spain, Sardinia, Tuscany, Naples, Brazils, West Indies, islands of the Pacific ocean, Cape of Good Hope, &c., must be sufficient evidence to convince any disinterested and unprejudiced mind of the utter folly of such an objection at the present day.

In a communication to the government of Norway and Sweden in 1830, M. Fresnel remarks upon this subject: "Happily, an experience of *seven years* has dissipated that fear, and the lenticular lights have been distinguished up to this time by the regularity of their service." Again, in reference to the same subject, M. Fresnel remarks, in a note to the undersigned, that "opinions thus expressed *fifteen years since*, based upon an experience of seven years, have been greatly strengthened up to the present time, embracing a period of *twenty-two years* since the establishment of the Cordouan light, and sustained by the results daily offered of more than one hundred and ten lights of the first three orders, established along the coasts of France and different foreign powers." "In this important point of view, then, the question seems to be irrevocably settled, and I will only add a few considerations relative to the application, more or less extended, which may be made of the new system of illumination to the vast maritime coasts of the United States."

It has been further objected that competent persons could not be procured in the United States to take charge of the lights fitted with the dioptric apparatus and mechanical lamps, for the salaries at present paid to light-keepers of the existing lights. The number of keepers necessary for those lights has also been urged as an objection to their introduction; and there is also a third objection, emanating from the same source, that the mechanical lamps could not be repaired when employed at distant or isolated points on the coast.

With regard to the keepers, no better evidence can be adduced than the opinions of M. Fresnel upon the subject, and the practical results furnished daily wherever the lights are employed. M. Fresnel says, "that the difficulty of obtaining proper persons to fill these subaltern stations appears to be most singularly exaggerated. "In France they belong almost always to the class of *ordinary mechanics or laborers*, who make from one and half to two and half francs per day (from 27 to 46 cents.)" "Eight or ten days will suffice ordinarily to instruct a light-house keeper in the most essential parts of his duty, receiving lessons from an instructor conversant with all the details of the service; and two instructing officers will be sufficient to prepare keepers for all the lenticular lights which could be successively established upon the coasts of North America." "In defence of this assertion, I will cite the example of the administration of Norway and Sweden."

As to the number of keepers allowed to the dioptric lights, there might be some reason in the objection, if it were possible to produce a light with parabolic reflectors possessing in any reasonable degree the advantages arising from the employment of a first order catadioptric apparatus; but as it is well established that reflectors are not susceptible (practically) of any combination which would produce a light equal in every respect to a first order dioptric light, the objection ought in honesty to be abandoned or waived by them, without they prefer bad to good lights, to guide the mariner in his perilous way along our shores.

The lower orders of dioptric apparatus, illuminated by ordinary Argand lamps and burners, with single and double wicks, require but one keeper, and they produce a light far superior to those of the same class in the catoptric system, independently of the economy in the use of the dioptric lights. In Scotland and in England, where the lights are as well if not better attended than in any other parts of the world, the same number of keepers are allowed for the same class of lights, without regard to the apparatus employed, whether catoptric or dioptric. At the South Foreland, for example, there are only three keepers for a first order dioptric and a first order reflector light, placed about three hundred yards apart, and at St. Catherine's a first order dioptric light has but two keepers to attend it; besides, other instances might be cited, if it were deemed at all necessary. But to accomplish in the most perfect manner possible the great and important objects for which lights are established upon sea coasts, it would seem but reasonable, and certainly desirable, rather to increase the number of keepers ordinarily allowed to catoptric lights than to diminish the number (taking France as a basis) for those fitted with dioptric apparatus.

In regard to the repairing of the mechanical lamps, it may be asserted, without the fear of being controverted, that in consequence of the superior manner in which these lamps are at present constructed in Paris, they will perform well for a number of years by bestowing upon them only the ordinary attention necessary to keep them clean ; besides, the number supplied to each light-house (from three to four, and never less than three) is a sufficient guarantee against any accidents which could prevent the proper exhibition of the lights. The same objections might, with equal propriety, be urged against revolving, flashing, or any other lights requiring clock machinery; yet such lights are found on every coast where lights exist to any extent. A simple inspection of the works of a mechanical lamp will convince any person of common understanding, that any mechanic who is capable of repairing the machinery for a revolving light is equally competent to put in order any lamp used in light-houses, and particularly those known as mechanical lamps with concentric wicks.

The oil of colza is used exclusively in the French light-houses. M. Fresnel says : "From numerous experiments it seems to me that these two oils (spermaceti and colza) may be employed with equal success *in lamps of single or multiple wicks.*"

M. Fresnel's preference for the colza (to the sperm oil) is based upon two reasons : first, the colza is less expensive in France than sperm, owing to the fact that the vegetable from which this oil is expressed, is cultivated on a very extended scale in France, Belgium, Holland, Holstein, &c.; and, secondly, the great difficulty in detecting impositions which may be and are practised by mixing inferior oils with the sperm, while, on the other hand, any impurities in the colza are very readily detected. No experiments have yet been made in France to test fully which of the two kinds of oil will produce the best light for light-house purposes.

 * * * * * * * *

There is but one floating light in France ; that is constructed of wood, moored and illuminated after the manner, with a few exceptions, of those belonging to the Trinity Board in England. The exceptions are: first, bronze is used in the construction of the lantern in the place of iron ; and, secondly, the lamps are mechanical, the pumps of which are put in play by springs instead of the ordinary fountain lamp. This latter, in spite of the delicate machinery of the lamp, is deemed a very decided improvement, as fulfilling much more fully the requirements of such a lamp, by preserving the centre of gravity in the same vertical during the whole time of the combustion.

REPORT OF THE TRINITY HOUSE ON THE RELATIVE POWER OF FRESNEL'S AND THE REFLECTIVE SYSTEMS OF LIGHTING LIGHT-HOUSES.

TRINITY HOUSE, *London, E. C., March* 5, 1857.

SIR : Having laid before the Elder Brethren your letter of 18th ultimo, signifying the request of the Lords of the Committee of Privy Council for Trade to be furnished with the results of any experiments which may have been made by this corporation on the relative power of Fresnel's and the Reflective Systems of lighting light-houses. I am directed to state, for their Lordships' information, that the reports which have from time to time been made to the Board by visiting committees in allusion to the comparative effect of the reflective and refractive principles when the juxtaposition of light-houses, illuminated therein respectively, have afforded opportunities of comparison, lead to the following conclusions, viz :

That, although the strength or power of the reflected light, when seen in the exact focus, may be considered at least equal, if not superior, to the refracted, the latter possesses the advantage of a more equal and uniform distribution of light over the whole circle of observation, and excels the reflected light when seen from positions which do not present the advantage of full focus; it may consequently be considered that the reflecting system is fully as efficient, if not superior, to the refracting for revolving lights ; and that the latter is superior for those that are fixed.

The consideration, however, which has induced the Elder Brethren lately to adopt as a general practice the use of the refractive apparatus is that of its comparative economy, for although the original cost of a first class apparatus on that principle considerably exceeds that of a reflective apparatus of the same class, the consumption of oil and the consequent expense of exhibition are materially less, and the apparatus being altogether of a more durable character, the cost of repair and renovation is also smaller.

The reflectors to be fully efficient for their purpose, must be kept in the highest possible state of brilliancy, and being liable to become dull or tarnished by the vapor which frequently arises in the lantern, the constant application of friction to their silvered surface is rendered necessary, which not only requires the strictest attention on the part of the light-keepers, but causes the gradual abrasure of the silver and the consequent necessity for repair or renovation.

Their Lordships are aware that a refractive apparatus on the holophotal principle has been contracted for by Messrs. Chance, of Birmingham, for the Lundy Light, but which being yet incomplete, the Elder Brethren are unable to offer an opinion from practical observation upon its alleged superiority over the refractive apparatus now in general use.

I have the honor to be, sir,

Your most obedient servant,

P. H. BERTHON.

The Secretary Marine Department,
Board of Trade.

REPORT OF MESSRS. STEVENSON, THE ENGINEERS OF THE COMMISSIONERS OF NORTHERN LIGHT-HOUSES, ON THE COMPARATIVE ELIGIBILITY OF THE CATOPTRIC AND DIOPTRIC SYSTEMS OF ILLUMINATION OF LIGHT-HOUSES.

In compliance with instructions from the Commissioners, we now beg leave to report our opinion as to the comparative merits of the catoptric and dioptric systems of illuminating light-houses; and we begin by stating the following principles, which will be useful in arriving at a correct conclusion:

We believe that optical arrangement to be the most perfect—

First. Which employs optical agents, consisting of materials that absorb the smallest number of rays, and produce the smallest amount of irregular scattering of the rays.

Second. Which sends the greatest number of rays to the eye of the most distant observer, provided the light is allowed to remain sufficiently long in view to answer the purposes of the navigator.

Third. Which produces the desired effect with the smallest possible number of optical agents; and

Fourth. Which has no unnecessary divergence, but which illuminates that arc only which is absolutely required to be illuminated.

If these principles correctly and fully embrace all the essential elements that must enter into any comparative view of the different systems of light-house illumination, it is obvious that, in order to discover whether the catoptric or dioptric is the better system, we must consider, first, what is the best material to be used in the construction of the apparatus; and, second, what is the best arrangement of the optical agents which are employed.

We may also premise, in order to prevent that ambiguity which has gradually crept in from a misappropriation of certain terms, that, first, by the term "*Catoptric*" we wish to include all optical reflecting apparatus in which the reflection is produced by *metallic* surfaces only; second, by the term "*Dioptric*" we include all optical apparatus which consist of *glass* only, whether acting by refraction only, or by refraction and "*total*," or, as it has been sometimes termed, "internal" reflection; and, third, by "*Catadioptric*" we include any combination of a metallic with a glass optical agent, or, in other words, any union between the catoptric and dioptric systems as above explained.

THE MATERIALS OF WHICH THE DIFFERENT KINDS OF LIGHT-HOUSE APPARATUS CONSIST.

In the catoptric system of illumination plated copper has long had the pre-eminence as the best material for reflectors, while in the dioptric glass is employed. In *Comparative absorption of rays by glass and metal.* contrasting the relative advantages of metal and glass, it must be observed that the loss of light in metallic reflection proceeds partly from irregular scattering of the rays, owing to imperfections in the form of the mirror and partly from actual absorption of the rays, which last depends mainly on the state of the polish. When rays of light fall on a metallic mirror, a very considerable portion of them are understood to suffer refraction, and, after entering the nearly opaque metal, they are believed to be finally extinguished or annihilated. In glass, on the other hand, there is a loss of light at each refracting surface, and also a loss by absorption, varying with the thickness and color of the material. Where the light is both refracted and reflected by prisms of glass, the reflection takes place where the rays impinge at an angle with the internal surface, less than what is termed the *critical* angle, where refraction becomes impossible, and, theoretically at least, there should not be a single ray of such reflected light lost, and hence it is called *total reflection*. It would be tedious to quote extracts from writers as to the advantages of employing glass rather than metal for altering the direction of the rays of light. We may mention that Sir Isaac Newton, Sir William Herschel, Sir John Herschel, Sir David Brewster, and others, have stated their opinions strongly on this subject. Sir John Herschel, for example, in the article "Light," in the "Encyclopædia

Metropolitana," says : "The reflection thus obtained" (viz: total from glass) " far surpasses in brilliancy what can be obtained by other means, from quicksilver, for instance, or from the most highly polished metals" (p. 369.) Sir David Brewster states ("Phil. Journal," 1832, p. 439:)—"I believe, however, on the authority of the phenomena of elliptical polarization, that in silver nearly one-half of the reflected light has entered the metal, and in other metals a less portion, so that we may consider the surface of every metal as transparent to a certain depth—a fact which is proved also by the transparency of gold and silver leaf.   *   *   *   It is well known that silver, polished by hammering, acts differently upon light from silver that has received a specular polish, and I have elsewhere expressed the opinion that a parabolic reflector of silvered copper, polished by hammering, will, from the difference of density of different parts of the reflecting film, produce at the distance of many miles a perceptible scattering of the reflected rays similar to what takes place in a transparent fluid or solid or gaseous medium."

Without quoting other authorities we shall refer to the experiments of Professor Potter, who has bestowed much attention to the subject of photometry, and whose experiments have in some respects differed from those of Bouguer and Sir William Herschel, and have rather tended to decrease the supposed differences between the light lost by metallic reflection and that lost by transmission through glass. He states, however, in his "Treatise on Optics," published in 1851, that only about $\frac{1}{30}$th of the light is reflected when light falls perpendicularly on a surface of common glass, which, if we suppose a similar loss at the second surface, would leave $\frac{28}{30}$ths to be transmitted, were it not for absorption due to the thickness and color of the glass, while he states, in the same work, that about one-third of the light is lost in a perpendicular reflection from ordinary silvered looking-glass, and a little less from highly polished speculum metal. The following tables show the quantity of light lost in reflection from metal, and in transmission through glass at different angles of incidence, deduced from Professor Potter's experiments:

Table of Experiments on Metallic Reflection.		Table of Experiments on Loss of Light in Transmission through Plate Glass ⅛-inch in thickness.	
Angles of Incidence.	Quantity lost by Reflection, &c.	Angles of Incidence.	Loss in Transmission.
. . .	. . . .	0°	·086
10°	·314	10°	·092
20°	·305	20°	. . . .
30°	·334	30°	·094
40°	·332	40°	·106
50°	·346	50°	·125
60°	·351	60°	·161
70°	·349	70°	·254

Now, whatever loss there may be from the passage of the light through glass of greater thickness than that employed in these experiments, it is obvious that there can be no appreciable loss caused by the refractions at the angles of incidence on the first order glass prisms which are in use in light-houses, as these vary from 7° 30′ to 45°, and in the lens from 0° to under 30°, and therefore fall wholly within the limits of the above experiments, which, in addition to the loss by refractions, include that of the absorption due to a thickness of ⅛th inch.

With regard to the absorption due to the passage of light through a thick prism, we have only a few experiments of Professor Potter to refer to. In the Phil. Magazine for 1832 he first demonstrates *experimentally the important fact that, as had been generally supposed, there is no loss of light by total reflection.* The whole loss is therefore due to the two refracting surfaces, and to the absorption in passing through the glass. His experiments on the amount of these are as follows: the loss of light due to every cause in traversing a totally reflecting prism of glass of the thickness of 1·98 inch, averaged ·234, where unity represents the whole incident light. The amount of glass traversed in these experiments is very nearly the same as in the more modern light-house prisms, so that the above results become quite applicable to our purpose. We also made some observations on pieces of straight rectangular prisms of the same size and quality as is used in light-houses, and by means of Professor Potter's photo-meter we obtained results somewhat higher than those just given. The results will be found detailed in Appendix No. 1. It is merely necessary here to state that, with 31 experiments, and with four different observers, we got a mean value of ·805 for the amount of light transmitted. We are far, however, from wishing to place our results on an equality with those of Professor Potter, whose great experience and accuracy in photometry are so generally acknowl-edged, and we are therefore quite content to adopt the lower results which he has obtained. It will be seen from his experiments, which have already been quoted, that by reflection from highly polished specula, within the limits of from 10° to 70° of incidence, the loss of light varies from ·31 to ·35; while the loss by total reflection is only about ·23. Here then is a clear gain of one-tenth in favor of glass over metal; or if we took our own results, there would be a some-what greater gain. But we must now notice a highly important allowance which has to be made in this case. Professor Kelland has kindly informed us that Professor Potter considers the results of his experiments on metallic reflection "quite inapplicable to light-houses, as the polish in his mirrors was such that the surface could not be seen when held near the flame of a candle, but appeared like a hole in the side of a dark box. The polish of a light-house reflector," he adds, "could never be kept up to this state." It would, indeed, be impossible to give such a polish even to a *new* reflector. The manner in which specula are polished, and by which alone such a perfect surface is attainable, is entirely inapplicable to a soft metal like silver; and neither can such a method of working be employed for any curve which varies much from a circular segment.

It clearly appears, then, that whatever may be the comparative gain due to the employment of total over metallic reflection, and we think there is every reason to believe it large, at least enough has been said clearly to establish the superiority of glass over metal, as a material for constructing light-house apparatus, so long as the inci-dence of the rays is not too oblique. It must also be observed that the loss from absorption may be reduced almost indefinitely by increasing the number, and thus reducing the size of the prisms which are employed.

In Appendix No. II will be found a number of experiments which we have made on the reflecting power of silver plate, polished in the same manner as a light-house reflector. From these it will be seen that not much more than *one half* (·556) of the light incident was reflected, a result which tallies with the statement of Sir David Brewster to that effect in his "Treatise on Optics."* If we assume the result as correct, and compare it with Professor Potter's valuation of a totally reflecting prism, we shall have a gain of two-tenths (·210) in favor of glass over metal; or if we compare it with our own experiments, we should have a gain of almost one-fourth (·249) by employing glass.

Before leaving the subject of the material to be employed in light-house apparatus, we must yet notice some further advantages which are peculiar to glass.

Curves formed in glass certainly admit of much greater accuracy in form than those of metallic reflectors. In the one case the result is affected by a very gradual process of grinding by means of unerring machinery of a rigid and unalterable construction; while in the other case the result is attained by a comparatively rude tentative manual process, and subject therefore to all the imperfections to which such methods of working are obviously liable. The polish, too, on which so very much depends, is in the glass apparatus, given *once for all* by the finely constructed machinery of the manufacturer, whereas the metallic polish is constantly undergoing deterioration from atmospheric oxidation, and requires to have its brilliancy daily renewed by a succession of different light-keepers, from the less skillful of whom it may receive ineradicable scratches and permanent injuries. Thus while the glass never loses its correct form, the metallic polish may be deteriorated and the curve of the mirror may be materially altered, although the existence of injuries and changes of form may never be suspected.

It is still, however, a possible case that improvements may be made on metallic reflectors, or that other suitable metals or metallic combinations may be discovered which are capable of receiving a superior and more lasting polish than silver. But unless it can be shown that there is indeed a very great gain from the use of the metallic reflection, we shall still remain decidedly of opinion that, irrespective altogether of the superiority of glass viewed as a material,

---

* "The great value of such a mirror" (one acting by total reflection) "is that as the incident rays fall upon A. C." (the reflecting surface) "at an angle greater than that at which total reflection commences, *they will all suffer total reflection*, and not a ray will be lost. *Whereas in the best metallic speculum nearly half of the light is lost.*"—Treatise on Optics, by Sir D. Brewster. London, 1831.

it possesses other peculiar advantages which render it decidedly more eligible as an optical agent. This will appear more fully from a consideration of the next subject which we propose to discuss, viz : What is the best optical arrangement for light-house illumination?

OPTICAL ARRANGEMENT OF APPARATUS.

Fixed lights. It is obvious, as appears from our second general postulate (*vide* p. 26,) that no apparatus can be regarded as the best which suffers any portion of the diverging rays which proceed from a flame to escape into the atmosphere without being parallelized in the directions required by the mariner, for such rays are altogether lost. In Fresnel's fixed light apparatus, where the object is to *illuminate constantly the entire horizon*, the whole sphere of light is emitted parallel to the horizon ; and as the apparatus consists only of a cylindric refracting hoop, and totally reflecting prisms, it produces its effect by the simplest conceivable combination of the best optical agents ; we are therefore of opinion that it cannot be surpassed, and must be regarded as the *optimum* form for fixed lights in insular stations. In narrow Sounds, where it is desirable to alter the intensity of the light in different azimuths in proportion to the different distances to which the light requires to be seen, we consider the holophotal condensing apparatus to be the proper arrangement. This construction, as the Board is aware, is now being adopted in several of the Sound lights on the west of Scotland.

Revolving light fitted with several independent lamps and apparatus, mounted on a revolving frame. For a revolving light, it is desirable that only a small horizontal arc should be illuminated by each flash ; and it is obvious that that arrangement must be the best which, without employing unnecessary optical agents (by which, as has already been seen, much light is lost,) condenses the whole sphere of diverging rays which proceed from the flame into the required arc or arcs of illumination. The only arrangement with which we are acquainted that fulfils these conditions, is that which has been termed the holophotal, which was first adopted in a harbor light at Peterhead, in 1849.

Catoptric and catadioptric arrangements. In the catadioptric form, it parallelizes the *whole sphere* of diverging rays into *one* beam by means of a certain union of parabolic and spherical metallic mirrors and a lens. These three instruments must be so combined as in no way to interfere with the proper action of each other, and in such a manner that all shall have one common focus situate in the centre of the flame of the lamp. The parabolic reflectors, commonly used in light-houses, suffer not much short of one half of the whole sphere of diverging rays to escape uselessly past the lips of the reflector. In the reflector proposed by Mr. A. Gordon, in 1847, there is also no attempt made to parallelize the rays which are immediately in front of the flame, and the surface behind the parameter of that reflector is so exceedingly small, that one half of the whole sphere of rays seems to fall upon a surface nowhere more than about one inch and a half from the circumference of the flame. The effect of this arrangement, which in other respects is a step in the right direction, is, practically speaking, to throw away perhaps nearly one half of the whole sphere of rays. For, in addition to the light which is lost in front of the lamp, and independent altogether of the enormous aberrations which must result from the slightest imperfections in the original form of what is, in this instrument, a most critical part of the apparatus, the divergence produced by a flame of one inch in diameter is, with this instrument, no less than about sixty degrees. Now, even although it were desirable, which it never is, to illuminate so great a horizontal arc by each flash of a revolving light, it is perfectly obvious that there can be no possible use of any divergence in the *vertical plane above the horizon*. From this it will be seen that such an apparatus cannot be regarded as at all perfect. Mr. W. Barlow has proposed (Royal Society Transactions, 1837, p. 215) to increase the illuminating power of a reflector by placing a spherical reflector in front of the flame, so as to intercept the rays which would otherwise be lost, and to return them back again through the light itself. This ingenious plan, though otherwise largely economical of light, necessarily renders wholly useless that portion of the reflector which is opposite to it, and it also occasions the loss by absorption of a considerable portion of the front rays. A far more complete optical arrangement is that which was proposed so far back as 1812 by Sir David Brewster, and afterwards introduced by A. Fresnel in his revolving lights. By Sir David's plan, the whole sphere of rays was usefully employed, and the excessive amount of divergence to which we have just referred was avoided. It labors, however, under the serious disadvantage of employing an unnecessary optical agent, which is obviated by the holophotal arrangement.

The holophotal dioptric arrangement may also be adopted where independent burners are used. In this case the whole apparatus consists of refractors, and

Dioptric arrangement.

totally reflecting prisms of glass made of a peculiar form, which prisms were first used in light-houses on our recommendation at Horsburg revolving light in 1851. When it is necessary to establish lights at distant and inaccessible places, we consider the holophotal dioptric system, with independent burners, to be the most eligible, and we are at present getting apparatus on this principle for the Board of Works of Newfoundland. Glass is, however, generally adopted in first order apparatus, having a single central burner, with the frame work of glass revolving round it. We shall now, therefore, refer to that system.

In the dioptric system on the large scale there is generally, as has just been stated, only one central burner, and the light is condensed into such number of parallel beams as may be found most convenient by the action of apparatus revolving round the central flame. We have already shown that the holophotal arrangements are preferable in the smaller class of apparatus, and it will at once appear that the same must be the case on the large scale. The first-class dioptric holophotal apparatus, consisting of lenses and totally reflecting prisms, similar to those of Horsburg, already referred to, produces its effects by means of *a single optical agent*, which is not the case in any other arrangement. The revolving apparatus of Fresnel, for example, is far from being (like his fixed light) a perfect arrangement. In it we find the light which passes below the lenses distributed all round the horizon, instead of within the arc of useful illumination, and the light which passes above the lenses is subjected to the action of inclined mirrors and lenses, by which the illustrious inventor himself admits that there is a loss of about half of the light incident on that portion of the apparatus. In Fresnel's fixed light varied by flashes there is a similar employment of *two* optical agents, whereas there is but *one* in the holophotal system. In the beautifully finished first order fixed light varied by flashes which was exhibited in the Great Britain exhibition of 1851 as a patented invention, and which was of French manufacture, there was a similar unnecessary reduplication; large cylindric refracting panels being placed in front of the ordinary fixed light prisms. While in the corresponding holophotal arrangement, on the contrary, the same effect is produced by the action of a *single optical agent*.

Dioptric system with one central burner.

But it is well known that the gravest objection to the Disadvan- dioptric system of lights as originally constructed is the tage of diop- smallness of its divergence. The great annular lens has tric system. a divergence not exceeding about 5° 9', whereas it may be desirable that it should in certain cases have perhaps twice this divergence. In the flashing holophotal light of the first order, at North Ronaldshay, (which was the first apparatus of the kind constructed,) the horizontal divergence was accordingly increased by placing the panels of totally reflecting prisms with their axis 3° 30' on each side of that of the central lens, and this amount of divergence has been found to be amply sufficient. Were it considered necessary for any purpose, as, for instance, in a revolving light with a long period of intervening darkness, to increase the divergence still further, nothing can be easier than to adopt the spherico-cylindric form of lens recommended by Mr. Thomas Stevenson, in the Edinburgh Phil. Journal for 1856.*

That apparatus, then, is obviously the best which condenses the rays with the minimum amount of divergence in every plane, providing always that it is easy, as has been shown in the case of the dioptric apparatus, to increase by a slight alteration in the arrangement the natural divergence just so far only, and in such planes only as may be required. It is, therefore, only with apparatus such as has been described, possessing this limited divergence, that the light is prevented from being uselessly projected upwards towards the clouds.

Having thus endeavored to show the relative merits Useful effect of the different materials, and different forms of appa- of different ratus, we will turn for a moment to the experiments systems. which have been made with the view of comparing the useful effects of different apparatus. We need hardly advert to the striking advantage of Fresnel's fixed dioptric apparatus of glass over the former

---

* The apparent superiority of the metallic mirror in respect of divergence is, in reality, as has been already shown, a great evil, for although such an amount of divergence in the horizontal plane, or in the vertical plane *below* the horizon, may be very well, yet it is a positive loss in the vertical plane *above* the horizon. But in the dioptric system, if it were ever found necessary under some peculiar circumstances to increase the vertical divergence *below* the horizon, this could be easily done by giving the whole or part of the apparatus a slight dip forward. We have not, however, in our practice ever found it necessary to adopt this plan. Reflectors, on the other hand, cannot be dipped so as to save the light passing above the horizon, because that would entail the loss of the light passing below the horizon, which would be thrown close to the light-house tower, where it is not required.

system of reflectors, as we can scarcely imagine the existence of any
doubt now as to its superiority. We will, therefore, confine our
attention to the subject of revolving lights. The numerous experiments which were instituted many years ago by the Board of Northern Lights at Gullan proved that a first-order lens equalled in effect
that of from seven to nine reflectors. The reflectors which were
experimented upon were those which were then in general use by
the Board, and it may no doubt be objected that had those reflectors
been rendered holophotal the result would have been different. In
proof of this we may mention that in 1850 a brass holophotal reflector
was tried against a Northern Light silvered reflector at Gullan, and
was viewed by observers stationed at distances of seven and nine
miles, who were purposely kept in ignorance of the different arrangements, and in every instance the brass holophotal reflector had the
advantage, and *on one occasion during fog it only was visible.* But it
must be recollected that since the date of the Gullan experiments
the lenticular system has acquired at least a corresponding advantage
over the old revolving light of Fresnel, by the adaptation of the
holophotal arrangement to lights of the first order. We may, therefore, for want of other experiments, conclude (with the high probability of being far within the mark) that now, as formerly, the illuminating power of the most perfect kind of lenticular apparatus of
the first order, and the most perfect kind of parabolic reflector, are
in the ratio of at least eight to one. Now, however suitable in many
situations and for lights of subordinate importance the dia-catoptric
system may be, we are strongly of opinion that for great sea lights
the dioptric is the far preferable arrangement. The advantages of
the dioptric system will more plainly appear if we keep strictly in
view that the grand desideratum for sea lights is an augmentation of
power, so as to give the mariner as great an "offing" as possible.
Suppose, for example, we wish to produce by reflectors the effect of
eight first-order lenses, we should require to provide a lantern capable of accommodating from 56 to 72 reflectors, and if so great a number of independent reflectors is admitted to be all but impracticable,
and we fall back upon one central four wicked lamp surrounded by
tiers of metallic reflectors, we shall not have gained anything by the
change, for the loss of light between the interstices which separate
the different portions of the mirrors will be great, and the errors in
position due to the daily polishing of so great a system of mirrors,
and the amount of labor that will be required, are such as to render

it necessary to have recourse to silvered glass. If such were the case we should have the loss of light due to the metallic reflection, and also to two refractions at the surfaces of the glass. Arago found by actual trial that the useful effect of such a system of reflectors, when compared with the totally reflecting prisms for Skerryvore light-house, were as 1 to 1-61. M. Fresnel also found the effect of a metallic reflector was to that of a catadioptric apparatus of the same size in the ratio of about 1 to 2.

For first-class lights, then, the case seems to be very clear; and, in connexion with this part of the subject, we would refer to the Appendix No. III. for the extract from a letter addressed to the Hydrographer of the Admiralty, regarding the erection of great sea lights on a gigantic scale, for increasing the safety of over-sea vessels. Were the construction of such sea beacons resolved on, there can be no doubt that the only practicable manner (in the present state of our knowledge as to the sources of light) in which the necessary power could be secured, would be by means of the dioptric system.

It may not be out of place, as having an important bearing on the subject, to mention, in conclusion, that in France, Holland, and Spain, and we believe in most other commercial countries, the preference has been already given to the dioptric system. In particular, the American Government, so lately as 1852, after having devoted most careful attention to the whole subject, and after having accumulated from different countries and from various quarters a great amount of information, which has been published at full length, stated the following as the conclusion of their labors (page 13 of their Report): "That the Fresnel or lens system, modified in special cases by the holophotal apparatus of Mr. Thomas Stevenson, be adopted as the illuminating apparatus for the lights of the United States, to embrace all new lights now or hereafter authorized, and all lights requiring to be renovated either by reason of deficient power, or of defective apparatus."

(Signed)   D. & T. STEVENSON.

EDINBURGH, *May* 6, 1857.

APPENDIX.

No. I.

TABLE showing the AMOUNT of LIGHT AVAILABLE, after passing through a totally REFLECTING PRISM, as prepared for ordinary LIGHT-HOUSE purposes. (Angle of incidence on external surface = 39°; length of passage through glass = 2·6 inches.)

The results are expressed in a fractional form, direct light being assumed = 1.

AMOUNT OF AVAILABLE LIGHT.	REMARKS.
·824	Mean of 4 observations.
·809	" of 3 "
·873	" of 4 "
·804	" of 3 "
·869	" of 6 "
·812	" of 4 "
·734	" of 3 "
·716	" of 3 "
·805	General mean of 31 trials made by *four different observers.*

NOTE.—The last two sets of observations appear to be decidedly too low, but they are inserted, as no error in the adjustment of the apparatus was detected at the time.

No. II.

Table showing the Amount of Light available after Reflection from surfaces of silver, polished in the same manner as Light-house Reflectors. (Angle of incidence 45°.)

The results are given in a fractional form, direct ligh being = 1°.

AMOUNT OF AVAILABLE LIGHT.	REMARKS.
·520	Mean of 3 observations.
·501	" of 3 "
·620	" of 3 "
·676	" of 5 "
·644	" of 4 "
·560	" of 4 "
·730*	" of 4 "
·380*	" of 3 "
·517	" of 4 " With another mirror.
·496	" of 4 "
·472	" of 4 "
·556	General mean of 41 trials by *three different observers.*
·556	General mean excluding the observations marked with *.

Note.—These experiments were made with two pieces of hammered silver plate, made as flat as possible, and as the polish seemed to be equally good in each, the difference of effect must have been due to some slight curvature, probably in opposite directions. We believe, however, that the mean cannot be far from the truth.

ᵃ These experiments, which were made successively by the same observer, appear to be inaccurate, but the cause has escaped detection.

No. III.

Extract from a letter, dated December 27, 1855, *to Captain Washington, Hydrographer to the Admiralty, by Thomas Stevenson, C. E., in reference to the question of the characteristics for distinguishing Lights.*

* * * * "What we want is powerful apparatus, not intricate distinctions. To be enabled to see a light in a thick night, though it be only half a mile further than at present, may be of incalculable moment. If, therefore, we can increase the *power* of our lights so as to make them pierce the gloom but that fraction of a mile further than they do at present, we are moving in the right direction. On that small amount of extra offing hundreds of lives may depend. This subject has been much before my mind lately, and I am of opinon that there ought to be constructed perhaps three or four great *Ocean Lights*, situate on different points in England, Ireland, and Scotland, facing the Atlantic, and possessing optical powers which would greatly transcend anything as yet attempted, or than is indeed required in many ordinary cases, where a smaller offing is needed; although in all cases superior power constitutes in reality superior safety. Such lights as I have mentioned would form noble protections for homeward bound vessels by giving them longer offings. For such purposes I would propose of course revolving lights as being most powerful, and I would adopt for them first-order holophotal apparatus arranged in a manner not yet attempted, but which would enormously increase the effect, without necessarily involving a greater consumption of oil. I must, however, reserve troubling you with this scheme at present. I intend, however, to devote my first leisure time to this matter, when I shall probably take the liberty of seeking your advice on the subject."

LIGHT-HOUSE PAPERS.

PLAN

FOR

DISTINGUISHING SEACOAST AND OTHER LIGHTS,

BY OCCULTATIONS.

BY CHARLES BABBAGE, ESQ., &c., &c., &c.,
LONDON.

RESOLUTION OF THANKS TO CHARLES BABBAGE, ESQ.

Resolved, That the Light-house Board of the United States have received, with much gratification, and have examined with great interest, the plan of distinguishing seacoast and other lights by occultations, proposed by CHARLES BABBAGE, Esq., of London, and kindly communicated to them, and will use every endeavor to have a full trial made of the method, which, in their opinion, promises such great advantages to the navigation of the world.

Resolved, That the thanks of the Board are hereby tendered to CHARLES BABBAGE, Esq., for his communication, made on the invitation of one of its members.

Adopted unanimously, November 26th, 1851.

WM. BRANFORD SHUBRICK,
President.

THORNTON A. JENKINS, *Secretary.*

NOTES RESPECTING LIGHT-HOUSES,

BY CHARLES BABBAGE, ESQ.

The object of these notes is to point out certain improvements in the use of existing light-houses, by which it shall become almost impossible —

1st. To mistake any casual light, on shore or at sea, for a light-house.

2d. Ever to mistake *one* light-house for another.

The plan requires, in most instances, no change in the optical means at present used for condensing and directing the illumination of light-houses. It adds slightly to the facility of observing them at great distances; and, from its simplicity and generality, is equally adapted to the use of all countries. Revolving lights must become fixed ; but the mechanism already existing for their rotation may, with little alteration, be employed for the motions required by the new system.

The principle by which these objects are to be accomplished, is to —

Make each light-house repeat its own number continually during the whole time it is lighted.

This is accomplished by enclosing the upper part of the glass cylinder of the Argand burner by a thin tube of tin or brass, which, when made to descend slowly before the flame, and then allowed suddenly to start back, will cause an occultation and reappearance of the light.

The number belonging to a light-house may be thus indicated to distant vessels. Take as example 243 :

1. Let there be *two* occultations.
2. A short pause.
3. *Four* occultations.
4. A. short pause.
5. *Three* occultations.
6. A longer interval of time.

This system of occultations must be repeated all night by proper mechanism.

The rapidity of the occultations themselves, the length of the pauses between the units and the tens, and between the tens and hundreds, as well as the duration of the long interval of time which marks the termination of the number, must be made the subject of experiment.

A light has been already used as an illustration, in which the occultations occurred at intervals of one second ; the pause occupied four and the long interval ten seconds. The pause was thought to be unnecessarily long, and was diminished. Whatever may be the times ultimately adopted, the experiments already made render it improbable that the average time required by a light-house for repeating its number, should amount to one minute.

It is by no means necessary that the counting of the number of a light-house should commence with the digit which expresses hundreds. No greater amount of time would have elapsed, if, in the above instance, the observer had commenced with counting the unit's figure. It would then have read thus : (3 occultations,) long interval, (2 occultions,) pause, (4 occultions,) pause.

But, since the long interval denotes the commencement of a number, it is already apparent that the number of the light-house is 243 and not 324.

In order still further to prevent mistakes arising from an accidental error in counting the number of occultations, it will be convenient to establish another principle for the purpose of numbering the light-houses.

Light-houses must not be numbered in the order of their position. But every light-house must have such a number assigned to it, that no digit occurring in the number denoting the several light-houses nearest to it on either side shall have the same digit in the same place of figures.

If five adjacent light-houses were thus numbered : 361, 517, 243, 876 and 182, supposing a mistake to have occurred in the first time of counting 243, and that it had been reported to the master of the vessel as 253, he would immediately, on looking at his numerical list of light-houses, perceive that a mistake had been made in the middle figure, because in any general arrangement, the number 253 would have been assigned to some light-house on a coast very distant from that on which 243 was placed. In fact, two out of any three figures would always detect the error of the third.

The law of numbering just stated is sufficient for the present object. Probably a little inquiry might produce a still better arrangement.

Thus, occultations would distinguish every light-house from all casual lights, and their number would identify the light itself. The whole illuminating power would be always employed, undiminished by the interposition of colored glass. These lights would be more easily visible at a distance, because it is known that the eye perceives more readily a faint light which is intermittent than an equal light which is continuous.

OF HARBOR LIGHTS.

The same principle of numerical lights is equally applicable to light-houses which indicate harbors. Information, however, of another kind, is often requisite for vessels about to enter them. It is always desirable that the depth of water, either within the harbor or on the bar, should be known.

This may be effected most simply by allowing the occultations of white light to indicate the number of the light-house, and instead of having a *long interval* of white light between each repetition, let a colored glass be placed before the light, and a number of occultations be made equal to the number of feet of water existing at the time.

Thus a tidal-harbor light-house will continually repeat its own proper number in white light, followed by the number of feet of water on the bar in colored light. If it should be thought desirable, it would be easy to make the color of the light *blue* when the tide is rising, and green when it is falling.

The mechanism for harbor lights need not be complicated, and by means of a float might be made entirely self-acting. The weight necessary for making the occultations might even be wound up by the float itself.

Another great advantage of a float is, that the depth of water indicated will always be the real depth at the time. The computed depth often differs from the true depth, owing to the influence of storms, and other accidental causes.

Some additions to this mechanism would enable it to indicate the depth of water on the bar by day as well as by night.

OF FOG SIGNALS.

During the prevalence of fogs, the lights which ought to guide the seaman are often indistinctly seen, or entirely obscured, until he has approached too near the danger against which they were intended to warn him.

In cases of fog, light-ships and light-houses are in some instances provided with gongs and bells, which are then kept constantly sounding. It is unfortunate that the means of warning the seaman of his danger should extend to the shortest distance when that danger is most imminent. The lights usually employed are visible at a distance from six to thirty miles; but the sound of a gong or bell is heard at a comparatively very small distance.

When these instruments are heard they merely indicate danger, but not its exact nature. It might in some cases be of great importance that the gong or bell should indicate the number of the light-ship. This could be accomplished by a very trifling alteration in the mechanism. Instead of striking the instrument at fixed intervals, let there be pauses and a long interval between the number of strokes which successively represent the digits of the number of the light-ship, just in the same manner as has been proposed for light-houses.

A light-house or light-ship whose number is 243, would be thus indicated during fogs :

(2 blows on gong,) pause, (4 blows on gong.) pause, (3 blows on gong,) long interval.

The same mechanism which caused the occultations of the light, might produce the blows on the gong.

The preceding explanations are sufficient to show that each light-house or light-ship, by continually repeating its *own number*, might render any mistake of it for a different light very nearly impossible. The great principle on which the system rests, is *to give numerical expression to each light.* If it be not thought necessary to apply it to every light-house, the most important may be chosen for its application. The expense of the alteration, and the amount of danger incurred by a mistake, will furnish the ground of decision in each individual case.

In proposing, however, a new system which has extensive bearings on other questions connected with the safety of those who travel on the waters, it is desirable that a general and comprehensive view should be taken of such of its applications as the rapid advance in mechanical and chemical science justify us in supposing must take place in a few years.

However partially the system may be adopted at first, a judicious foresight into its probable applications may enable us, without any present inconvenience, to accelerate future improvements, and to save considerable expense on their adoption.

The following suggestions for improvements or applications, many of which are perfectly practicable at the present time, are offered for the consideration of those who may be called upon to carry out the *Numerical System of Light-houses*. They are not necessary for the success of the simple plan which has been already described, but may be adopted or rejected without any interference with it.

SUGGESTIONS FOR THE IMPROVEMENT OF LIGHT-HOUSE SIGNALS, BUOYS, &C.

Telegraphic communication during the night between light-houses and ships in distress.

Cases occur in which it is of great consequence that a ship should communicate with the land long before it can send a boat ashore, or enter its intended port. It may be the bearer of important intelligence. It may convey some personage whose presence is essential for some great object. The vessel itself may be in distress. The state of the elements may render it impossible to send for, or receive, any assistance from the land ; yet, even under such unfavorable circumstances, if directions from skilful pilots acquainted with the coast could be conveyed to the ship, its wreck might, perhaps, be prevented ; or, if driven on shore, having been directed to the least unfavorable spot, its crew might possibly be saved.

Such communications might easily be organized. There are already existing in the Royal Navy in the East India Company's service, and elsewhere, large dictionaries of numerical signals. These, it is true, are made by flags, or by balls ; but the same numbers may be expressed by the occultations of lamps. Any number, however large, may be expressed by making the number of occultations corresponding to the first or highest digit, then allowing a pause ; after which the number of occultations representing the second digit, then a pause; and so on, always observing that, after the unit's figure has been expressed, there must follow a long interval.

The plan for telegraphic communication would be thus arranged :

1. Light-house repeating its own number.

2. Ship fires a gun, and hoists a light, to call the attention of the light-keeper.

3. Light-house ceases repeating its number, and becomes a steady light, thus informing the ship that it is observed.

4. Ship having prepared its message, numerically expresses it by the occultations of its own lamp.

5. Light-house repeats the message of ship, in order to show that it has been rightly understood.

6. Light-house now repeats its *own number*, whilst it is preparing the answer.

7. Light-house expresses its answer by occultations.

8. Ship repeats the answer.

This interchange of question and answer is continued as long as necessary, during which the light-house repeats its own number previously to each reply.

Very little delay will occur, for these questions and answers will be arranged on movable disks, which may be placed in the mechanism employed for occulting, even whilst it is repeating another message. Many such disks, each containing a different message, may be placed in the machine at once, and on touching any lever the light will continue repeating the corresponding message.

In case of a ship in distress, for instance, requiring an anchor of given weight, it may be necessary to send to the harbormaster of the adjacent port to give the order, and to ascertain the time when it can reach the vessel. During this interval, the light-house will be repeating its own number.

An electric telegraph from the light-house to the dwelling of the harbormaster would save much time, and in some cases much damage.

The gun fired by the vessel might also be heard by the harbormaster, and his attention then being directed to the telegraph light-house, the whole time might be saved. If even his own house was invisible to the ship, but within view of the light-house, he might by means of a small light correspond with the ship through the intervention of the light-house, repeating the signals of both parties.

Colored shades might, if thought expedient, be used for different dictionaries ; or an entirely independent lanthorn might be specially

4

devoted to signals ; but this would cause additional expense, and seems unnecessary.

It may be objected to this plan, that it would mislead other vessels on first coming in sight of the light-house. This objection, however, will be found on examination to be invalid ; for a ship on first getting sight of a light-house will be at the distance of many miles ; and as all telegraphic messages would consist of more than three places of figures the ship would immediately perceive that the light-house was acting telegraphically, and on turning to the dic· tionary, would even become acquainted with its message. Besides, in the course of every three minutes, at least, the light-house would repeat its own number. Thus the ship would always know that it was in the presence of a light-house ; and if its reckoning did not enable it to identify the light, it could only remain in doubt during a few minutes.

Telegraphic signals between ships at night.

The application of the system of occultations to ships at sea may not, perhaps, be quite so easy as that which is proposed for light-houses, but no objections have yet occurred which appear at all insurmountable.

The question of the position of the occulted light or lights placed on the ship must be settled by practical men after due consideration and experiment. It may, however, be suggested that a light hid by a mast or sail may yet have its occultations made perfectly apparent by reflection from another sail. If such a system of signals were adopted, fleets might sail in company during the night, each repeating its own number, and any orders could be conveyed to any individual ship.

Specific lights have already been employed to distinguish sailing·vessels from steamers, in order to prevent collision. By adopting the system of occultations to one or more of the lights of steamers, their character would appear more distinctly, and at greater distances. Perhaps, indeed, it would be better to have the distinctive character of a steam vessel indicated by a continual enlargement and diminution of its light, rather than by an occultation. Two steamers also would have much less reason for approaching each other, because they could hold any correspondence by signals. They might also by the same means convey to each other their intended course long before they approach each other.

Of a universal dictionary of signals.

Whether the system of occultations be generally adopted or not, numerical dictionaries of signals have been found absolutely necessary, and have long been in use. The rapid increase both of ships and of steamers renders some common language for all nations almost a matter of necessity.

The concurrence between adjacent nations in numbering their respective light-houses would be essential if any *numerical system* is adopted for distinguishing them. Such an opportunity ought not to be lost of rendering those discussions still more useful by attempting to organize a plan for an universal system of numerical signals. The first step might, perhaps, be that each nation should supply all questions and answers that ships could ever require for their safety or convenience. Out of these, the duplicates being omitted, the first draught of the naval part of the dictionary might be formed. This being submitted to criticism would probably itself suggest many additions.

The questions should be very carefully translated into the languages of all maratime nations, and should be printed in columns for each language.

A dictionary of this kind, containing about five thousand terms, in ten European languages, was published in 1849, by M. K. P. Ter Reehorst. The words are contained on about two hundred double pages ; and since each word, of which there are usually about twenty-five in a page, is numbered, this work might be used as a numerical telegraphic dictionary.

If a more general dictionary were undertaken, other considerations arise, and the great questions relating to the philosophy of language must be examined with reference to such a work. It will, however, be sufficiently early to enter on that subject when any steps are seriously taken to accomplish so desirable an object.

The continually increasing use of the electric telegraph renders an universal language still more desirable.

ON THE IDENTIFICATION OF A LIGHT-HOUSE.

A case has been more than once suggested to the author, to which it may be desirable to advert in order to point out the course of experiment which may lead to its removal.

At certain periods of the year, and on certain coasts, there occur dense fogs. Under these circumstances it has happened that a vessel has, on a partial and momentary opening in the fog, insufficient to show more than a single occultation, found herself almost close upon a light-house. In such a case there is neither time nor opportunity to ascertain its number.

It may here be remarked, that the assumed danger of going ashore is so imminent that it is not *necessary* to know the number. It is sufficient for the moment to know that there is a light-house in a certain direction, which is close at hand.

It must, however, be admitted, that in common with all received systems of lights, the method of occultations will not furnish a remedy. If a colored light is already employed in particular localities to meet such a case, it will still accomplish the purpose when occultations are applied to it.

The danger, although rare, ought, however, to be provided against. The following remarks are suggested to assist in obtaining that object:

The time between two occultations (usually one second) might be doubled in special cases. A little experience would enable most men to recognise the fact after two occultations. If such light-houses were placed alternately with others, no light-house would be mistaken for either of its adjacent neighbors. This plan might be partially extended, but it is liable to objections.

Another view may be taken. Is it possible to give a specific character to the occultation itself? It has been found that if the occulting cylinder descend rather slowly over the lamp, and then, after a *very short* pause, rise suddenly, the effect is best. It has also been observed, when an accidental defect in the apparatus caused the cylinder, after suddenly rising up, to rebound, and again to obscure partially the lamp, that the nature of the occultation was peculiarly characteristic. This peculiarity was very remarkable up to a certain distance, after which it became lost. Almost any form of peculiarity can be given to the occultations, by giving proper forms to the cams which govern them. The fact that such peculiarities are not seen until the ship has approached within certain distances, does not appear to present a material difficulty, and may even prove an advantage.

It would seem, then, to be desirable to institute a series of experiments to determine the following questions : Can the occultations of a lamp, in which the rapid reappearance of the light occurs from the falling *down* of the shade, be distinguished from those in which it occurs in consequence of the rapid *rising up* of the shade ? and, if so, at what distance ? In some cases the shades might move from right to left, and in the reverse direction. What peculiarities in occultations can be seen at the greatest distances ?

Amongst the experiments still required, may be mentioned the loss of light resulting from the interposition of colored glasses, and also the proportion of light lost by sacrificing *given* portions of various parts of the optical apparatus used for concentrating it. This is necessary in order to enable us to judge what portion may be most economically sacrificed in case the space might be required for other purposes.

The dangers arising from fogs are of such an extent that all the resources of science ought to be called in to remove them.

Voltaic light can scarcely be depended upon, except under continual superintendence ; it would therefore be expensive. If, however, any intense light can be found capable of penetrating dense fogs, it might, during their continuance, be good economy to employ it even at considerable expense.

Perhaps the ordinary light-house lamps might be supplied with *oxygen* during fogs ; its expenditure being regulated by the obscurity to be penetrated.

Possibly, portions of phosphorus might be burnt in oxygen, and the light-house would then express its number by a series of *flashes*, and of *pauses* between them. The new form which that body is now known to assume might render its application to this purpose free from danger.

ON SOUNDS USED FOR SIGNALS.

Both gongs and bells are employed as substitutes for lights during fogs. I am not aware of any series of experiments on the distances at which sounds of various kinds can be heard. In a question on which so much property and so many lives depend, it is surely important to be well informed. The only resource is experiment. It may be remarked that the low notes of the gong might be confounded with those of the roll of waves breaking on the shore, whilst the shrill whistle of the steam engine will find a rival in the wind whistling through the rigging. The trumpet and the new and still more powerful instrument at the recent exposition ought also to be compared.

Again, although some of these may be heard at greater distances in the open air, some may be more easily adapted to have their sound concentrated and directed, when placed in the focus of a parabolic mirror, or, perhaps, at the end of a long tube.

Sound is transmitted to considerable distances through water, and it has been suggested that this might be used in case of fogs. But it seems probable that sound would be much interrupted in its progress, from the constant motion of the waves; and if it were transmitted at a considerable depth, it might be difficult for a vessel to send down an apparatus to render it sensible.

Experiments should be made on the distance at which sounds can be heard under water in various circumstances of its motion.

If, during storms, the surface only is agitated, it might be possible to transmit sounds in the still water near the bottom to considerable distances. Thus channels might be traversed by telegraphic communications with a less costly apparatus than that of the electric wire. It ought also to be ascertained whether the forms of the instruments struck would enable them to project their sounds in particular directions. Gongs, bells, and the firing of cannon under water, are among the sounds to be tried.

Whatever may be the sound audible at the greatest distance, it will be necessary to ascertain what are the best means of producing it in greatest intensity—whether by one large instrument, or by many small ones. It seems probable that some combination of discordant sounds may be most effective, because it seems to be a law of our nature that contrasts produce stronger impressions than uniformity. There is one form of sound the most disagreeable with which we are acquainted; it is said "to set the teeth on edge." What is the cause of this, and does that highly obnoxious sound penetrate further than others? If it penetrates as far as others, it will certainly be the earliest to be noticed.

LIGHTS ON BUOYS.

The time is probably not remote when lights will be placed on floating buoys, for the purpose of pointing out isolated dangers—as sunken rocks, shoals, &c., on which light-houses cannot be placed, or where the great expense may prevent them from being built. They may also be useful to indicate the channels leading to some few ports of very great resort, in order to render the approach of vessels possible during the night.

The first difficulty in placing lights on buoys arises from the necessity of trimming the lamps, and of supplying them with fresh oil. Galvanic processes seem to present a similar difficulty. The chemical discoveries of recent times, however, offer some hope of removing it. By the destructive distillation of peat, of coal, and of shale, as well as by other methods, a variety of combinations of hydrogen and carbon have been obtained. Some of these only remain liquid under a pressure of two or three atmospheres. They possess considerable illuminating power; and by confining them in a close vessel, and allowing a very small aperture for their escape in the state of gas, a jet of flame may be produced, of uniform magnitude, and without the use of a wick, until the last drop of fluid has evaporated. If such a fluid could be produced at a moderate price, a quantity might be enclosed within the buoy, sufficient to last several weeks, if not months.

Such a light would burn without the necessity of trimming, but it would require mechanism to light it each evening, and to put it out each morning.

Such mechanism already exists in many of our public clocks. If it is thought desirable, too, that it should occult, so as to indicate its number, the plan already described might be applied. Thus the buoy would contain two pieces of mechanism. The only remaining difficulty would be the necessity of visiting the light frequently in order to wind up the two instruments. This might probably be removed by having within the buoy a heavy pendulum, or perhaps two such, swinging at right-angles to each other. If the *perpendicular* motion of the buoy could be secured, then the winding up pendulums must be maintained horizontally by means of a powerful spring. These, by the action of the waves, would be continually winding up the springs which drive the mechanisms. This might be so arranged that it would never over-wind them.

Spirits of turpentine, benzole, and several other compounds. assume a gaseous state at very low temperature. If the end of a tolerably thick rod of metal is heated by the flame of the lamp, and the other end conducts the heat to the bottom of the fluid, it is sufficient to produce a continuous stream of gas to supply the burner until the last drop of the fluid is exhausted. Lamps constructed on this principle have, under various names, been in use for several years. If the fluid were sufficiently cheap, one of these movements might be dispensed with, by allowing the light to burn constantly during the day as well as the night.

New forms would be required for such buoys. Probably a columnar form, weighted at the bottom, might give a steadier light amid the fluctuations caused by the waves. These buoys should be attached to their moorings by rings fixed at the centre of resistance.

OF THE MECHANISM NECESSARY FOR OCCULTING LIGHTS.

The period of time occupied by any occulting light in making a signal is so short that great accuracy in the wheel-work is not necessary. In light-houses the moving power may be a heavy weight driving a train of wheels. This must terminate in a governor, which presses by springs against the inner side of a hollow cylinder.

When the length of the time necessary to indicate the number of the light-house is known. the governor must be so adjusted that some one axis shall revolve in the given time. A cam-wheel must be fixed on this axis, having its cams and blank spaces so arranged as to lift up the tail of a lever carrying the occulting cylinder at the proper intervals of time. Each tooth of the cam-wheel will cause an occultation of the lamp by the cylinder, which is instantly drawn back by a spring.

It is obvious that an axis might be used which moves round in the course of two, three, or more cycles. In this case the same system of cams would be repeated an equal number of times in the circumference of the cam-wheel. This plan is sufficient for light-houses which are not intended for signal stations also.

When signals are to be used, it is better to have a single cam on an axis which revolves once in the time which elapses from the end of one occultation to the end of the next. The effect of this cam will be, by acting upon a forked lever, to lift up the occulting cylinder. If nothing retain it in that position, the action of the spring on the lever will cause it to descend, and the cylinder, acted on by gravity, will instantly follow. But if an arm is interposed which retains the cylinder, then the forked lever alone will be pulled back by its spring, and the occulting cylinder will remain suspended until the next turn of the cam-wheel.

The suspending arm which was interposed must itself be governed by a cam-wheel, expressing the number of the light-house.

When a signal is to be made, an adjustable cam-wheel is to be set to the proposed signal, and is to be fixed upon the axis carrying the constant number of the light-house. When the proper time arrives for making the signal, it is only necessary to shift the axis, so that the adjustable cam-wheel shall be moved into the place occupied by the fixed cam-wheel. The signal will now be made and repeated as often as required, after which the original position of the constant cam-wheel must be restored. It is clear that any number of adjusting cam-wheels might be prepared for signals, and put upon the axis at once, so that a series of different signals might be made in a very short time.

Lights to mark the depth of water must have a heavy float connected with them, which at every foot of its rise or fall, must alter the number of occultations made by the colored light. It must, also, at the turn of the tide change the color of the light. It is sufficient for the present purpose to observe that the mechanism similar to that by which a clock strikes different hours, might be employed for this purpose.

The well in which the float is placed ought to be open to the tide by several small apertures; this would render the rise or fall of the float more uniform.

Telescopes are used for observing light-houses. They have a small magnifying power, but a large aperture. It is important that they should be as short as possible for taking in a given visual angle. Possibly, those constructed with a lens of rock-crystal might be employed with advantage, but upon this subject, also, experiment must be made.

PARIS, *July* 29, 1851.

SIR : The letter which you have done me the honor to address to me under date of 27th of May last, only reached me on the 5th instant, and unfortunately at the very moment of my departure for London, so that I was compelled to defer replying to it until my return to Paris.

I still belong to our light-house commission, but simply as a member having a deliberative voice. I obtained, some five years since, permission to retire from active service in consequence of declining health, and was replaced in the double capacity of secretary to the commission and inspector of light-houses, by my fellow-laborer of long-standing, M. L. Reynaud, engineer-in-chief, professor of architecture in the Polytechnic school, and in the school of bridges and roads. (Ponts et Chaussées.)

Consequently your letter should have been addressed to Mr. Reynaud, and you had better place yourself in relations with him for all information concerning our light-houses ; information which he is now more able to furnish than myself.

I have, therefore, thought it best to return to the legation of the United States the documents transmitted through it from yourself, with a request that they may be addressed to our minister of public works. This high functionary, I am well assured, will most cordially call the attention of M. Reynaud to the wishes of the American commission ; and you may rely with certainty that this learned and skilful engineer will coöperate zealously in promoting the success of the important vocation which has been entrusted to you.

Nevertheless, in order to fulfil, so far as I can, the expectations of the eminent persons composing your Light-house Board, and to obviate as much as possible the consequences of an unavoidable delay, which I regret not being able to prevent, I will endeavor to answer immediately some of the questions you have addressed to me.

I have just read over the minutes of *two notes* which I had the honor to transmit to you under the dates of 31st of December, 1845, and 13th of January, 1846 ; and I can but reassert the declarations and observations contained therein.

Theoretically considered, the relative merits of the two systems of illuminating—catoptric and dioptric—seem to me to have long since been clearly established.

In a *practical* point of view, the question does not, under all circumstances, present itself in so simple a form. There are, I am aware, certain localities, which, in consequence of the difficulty of communicating with them, together with their distances from workshops, would cause some hesitation in confiding to keepers the charge of mechanical apparatus, when so isolated from the means of repairing those accidental injuries which might derange the whole service of seacoast illumination.

Nevertheless, it is to be observed :

1. Wherever there is no hesitation in using the *revolving apparatus*, there should be none in the application of mechanical lamps ; particularly with the precaution of having always two or three spare ones.

2. After all, the difficulty in such cases resolves itself almost in every instance into a question of expense, which is a minor consideration where an important light-house is in question. In fact, in order to secure every precaution, it may be sufficient to secure for the chief keeper a mechanic, provided with a small number of tools and other necessaries, who will himself repair any accident which may happen, but which in reality very rarely occurs.

3. In short, communication by steam in the United States is now so wonderfully developed that no point in its immense coast can well be considered as isolated.

As regards the feasibility of converting a catoptric light in full operation into a dioptric one, it is a question for special consideration, requiring in its solution a very searching examination into local circumstances, and into the means of supplying a temporary illumination during the progress of transformation. If it be determined to erect a new tower near the old one, of course all difficulties are removed. The operation becomes a more delicate one if the new apparatus is to replace the old one in the same edifice. Nevertheless, a similar operation has been very successfully effected in our light-houses of Cordouan, Ushant, Du Tour, and Havre, (2 ;) also in the English light-houses of Inchkeith island, the Isle of May, South Foreland, Eddystone, &c., &c.

In France we have not only substituted almost entirely the catoptric apparatus of our old light-houses and beacons by the dioptric one, but we purpose in a short time to renew two or three of our oldest lenticular apparatus, not that they have in any manner deteriorated, but because their plan of construction has been so much improved upon by the skill of our artisans of the present day.

On the 1st of January of this year the illumination of the coasts of France and her colonies comprised one hundred and ninety-eight lights, large and small, out of which one hundred and thirty-two had the lenticular apparatus; thirty-one of which were of first order, six of the second, eighteen of the third, and seventy-three of the fourth. The increase since the 31st of December, 1845, has been, in lenticular apparatus twenty-one, of which four were of the first order, three of the second, and fourteen of the fourth.

Our seacoast illumination being about completed, at all events for the coasts of France and Corsica, will receive little or no modifications, so far as the lights of the three first orders are concerned. As to the small movable beacon lights, fixed upon scaffolding at the entrance of several of our harbors, they continue to be lighted by small reflectors.

I have not the documents necessary to enable me to reply to your questions concerning the *foreign lenticular lights.* I only know their numbers to be already considerable, and that fresh orders have reached our artisans from Denmark, Sweden, England, Spain, and Italy.

Some lenticular apparatus have been made in England by Mr. Cookson. of Newcastle, who soon renounced a manufacture which he doubtless found of but little profit. Two lenticular apparatus of the first order, one English, the other French, are now displayed at the great exhibition in London. So far as the cutting of the glass is concerned, I conceive them to be of equal merit; but in whiteness, the Birmingham glass cannot stand a comparison with that of St. Gobain. As to what relates to the organization of the light-house service, the most important modification since my retirement appears to me to have been the applying to our whole seacoast the system of administration which we had previously tested for several years on the coast of the English channel. By this measure all the light-keepers have been placed under the exclusive direction of the engineers, so that we shall have no longer the struggle against the parsimonious tendencies of the contractors.

I subjoin, herewith, two documents, drawn up and published by M. L. Reynaud in 1848, to wit:

1. New directions for the keeping of lenticular lights.

2. A schedule of charges for the furnishing of oil for the illumination of our coasts.

I shall close here, my reply, which may cover, satisfactorily, the principal points in the series of questions you have presented. More ample information will doubtless be forwarded to you by M. Reynaud, especially upon the improvements introduced into the construction of lenticular apparatus, with their lamps, upon the present cost of these apparatus, and upon the number sent from Paris to foreign countries.

You may also consult with advantage two works published by Mr. A. Stevenson:

1. Account of Skerryvore Light-house, in folio, Edinburgh, Adam and Charles Black—1848.

2. Rudimentary Treatise on Light-houses, in 12mo.; London, John Weale, High Holborn—1850.

While regretting that my present position has not enabled me to be more full and satisfactory, I can but congratulate myself, sir, upon the occasion afforded me of recalling myself to your kind recollection, and renewing the assurances of the high esteem of your devoted servant, **LOR. FRESNEL,**

Divisionary Inspector of Bridges and Roads,
and Member of the Light-house Commission,
No. 52 Rue de Lille, à Paris.

Lieut. THORNTON A. JENKINS, U. S. N.,
Secretary Light-house Board, Washington.

LIGHT-HOUSE COMMISSION,
Paris, July 23, 1851.

SIR: I have the honor, herewith, to address you a note in reply to the questions contained in your letter of 27th May last, which only reached me on the 7th of the present month. I send you at the same time a collection of documents (printed) relating to the light-house service, also some specimens of wicks and glass chimneys, such as are now in use.

I regret that the business with which I am overwhelmed on the eve of a tour of inspection, involving an absence from Paris of three months, will not allow me to enter more into detail in my answers; but I know that my very honorable predecessor, Mr. Fresnel, has already supplied you with documents on this subject, and that your enlightened and perspicuous mind will well admit of conciseness.

You will find me, however, very ready to reply to all calls upon our experience which you may do me the honor to address to me.

Be pleased, sir, to accept the expression of my high consideration,

L. REYNAUD,

Engineer-in-chief. Secretary to the Light-house Commission, charged with the direction of the service.

To Mr. JENKINS,
Lieut. U. S. N., Secretary Light-house Board of the U. S.

A reply to the questions addressed to the engineer-in-chief, secretary to the Light-house Commission of France, by Mr. T. A. Jenkins, secretary of the Light-house Board of the United States.

Mr. Jenkins, secretary of the Light-house Board of the United States, has done me the honor to address to me, in the name of this Commission, several questions relative to the light-house service. I will reply to them in succession, following the same order in which they have been put.

1. What have been the increase and improvements in the seacoast illumination of France since the year 1845?

The reply to this question is shown by the comparison between the annexed list of light-houses illuminated on the 1st January, 1851, and that which was published in 1845. The following table will exhibit this more plainly :

Apparatus.	No. in 1845.	No. in 1851.
LIGHTS OF FIRST ORDER.		
Lenticular ·	31	*35
Reflector ·	4	†2
LIGHTS OF SECOND ORDER.		
Lenticular ·	3	5
Reflector ·	1	0
LIGHTS OF THIRD ORDER.		
Lenticular ·	13	16
LIGHTS OF FOURTH ORDER—BEACONS.		
Lenticular ·	57	66
Reflectors · · · · · · · · · · ⸲ · · · · · · · · · ·	42	40

° The lights of Algiers and the colonies are not comprised in this table. The two lights of the first order at the mouth of the Canche are not included in this table. It is expected that they will be lighted in a few months.

† These two apparatus will, in all probability, be replaced in 1852 by lenticular apparatus.

2. What improvements have been introduced into the Carcel lamps ; and are other lamps used?

According to the annexed statement of instructions, published in 1848, upon the service of lenticular lights, three kinds of mechanical lamps were in use at that time in the lighting of light-houses of the three first orders. They are in use still. Of these, the Wagner lamp, the last of the three in the order of invention, is perhaps the one most to be relied upon for regular use ; its mechanism is very simple, and it is secured with all necessary solidity. There has just been made, for one of the lights of the first order, which will be shortly lighted at the mouth of the Canche, a lamp with four concentric wicks, placed according to the system of the moderator lamp, except that the spiral spring is replaced by a weight which acts on the piston, (see description of the moderator lamp, p. 8 of the instructions upon the service of beacons.) This lamp—which would be less expensive than the preceding ones, and more easily kept—has given very good results in the many trials which have been made in the light-house workshops ; nevertheless the result of a longer trial will be waited for before applying it to other light-houses.

3. What oil is used?

Colza oil is used in the light-house service. Numerous experiments have proved that this oil is the best for lighting ; it is, moreover, at present the most economical. Some years since the contractors for the furnishing of oil asked authority to substitute for it olive oil, for the lights of Corsica, on account of the difficulty of procuring the oil of colza. They stated, besides, that the first of these oils, properly prepared, was eminently suitable for lighting.

Experiments were then made to compare the two oils. Two lamps of the first order, similarly placed, were filled, one with colza oil, the other with olive oil ; they were lighted, and the flame received, as nearly as possible, the same degree of adjustment. Two metallic screens, each pierced with an opening of the same dimensions, were placed before the flames, and at the same height, and the rays of light passing through these openings were measured. This experiment was repeated several times, and it established that the light of the flame from the oil of colza was represented by 1.00 ; that of the other flame had for its mean value 0.88.

It was proved, besides, that the consumption by the two flames, at the same graduation, was represented, for the colza oil, 1.00 : while that of the olive oil reached 1.137.

The authority asked for was refused.

Other experiments were made, a few months since, on a mineral oil—the oil of schiste. They established that, when the flame was well managed, it gave a light notably superior to that which is obtained from the oil of colza ; but when the flame was agitated, or the lamp did not perform perfectly well, it evolved extremely thick smoke, which quickly darkened the apparatus. It would be necessary to have a lamp specially adapted to the burning of this oil. These experiments will be followed up, and they may perhaps lead to satisfactory results ; but in the actual state of things the oil of schiste cannot be applied to light-house illumination.

Some trials have been made with hydrogen gas, and with a mixture of hydrogen and oxygen, and they have given good results as to the quantity of light produced, but the expense would be greater : and, what would be a capital defect, less certainty in the continuance of the light.

4. To what tests are the oils, wicks, and chimneys submitted?

The articles 8, 9, 10, 11, and 12, of the schedule of charges relative to the furnishing of colza oil, will answer the first part of this question.

As to the wicks and chimneys, they are examined to see if those delivered are conformable to the model deposited in the central workshop for light-houses. There is sent with this note specimens of the different kinds of wicks and chimneys, which, after numerous trials have been definitely adopted for the service of lights.

5. At what periods are the inspections of lights made?

The light-houses are placed in charge of the engineer corps of bridges and roads, (*ponts et chaussees.*) In each district there is an engineer-in-chief and engineers; these last have under their orders agents who have the title of conductors. Each light is placed under the immediate superintendence of a conductor, who visits it at least once a month, but in most cases oftener.

The engineer never remains more than three months without visiting the lights of his district, and the engineer-in-chief inspects those of his department at least once a year. Finally, the engineer charged with the direction of the light-house service makes each year an inspection, which embraces one-third of the sea coast, in order that he may examine each of these establishments at least once during the three years. These visits are made by day and by night. Each conductor is provided with keys by which he may enter unawares, even into the lantern itself, in order to ascertain if the light be attended to according to the regulations. The port-officers and harbormasters have orders to transmit to the engineers all complaints from navigators, upon the keeping of the lights.

6. What is the number of keepers to each light?

The number of keepers is fixed at three for lights of the first order, and two for lights of the second and third orders. Beacon-lights have but one keeper; but another keeper is added when these establishments are placed on rocks isolated in the ocean, or uninhabited islands, that these agents may have by turns a regular leave of absence, without exposing the service to suffer therefrom. The detailed estimates of the annual expenses of illumination will show the number of keepers in each of the light-houses in France.

7. What is the annual expense of repairs to the Carcel lamp?

The detailed estimates, which served in 1838 as the basis of the contract for illuminating the light-houses of the ocean and of the Mediterranean, had valued this expense at seventy-five francs per year for the lights of the first order, at sixty-five francs for those of the second order, and at forty-eight francs for those of the third order. This valuation was too large. Thus, in 1850, the expense incidental to this object for twenty-eight lights of the first order, four of the second order, and thirteen of the third order, only reached 917.80 francs, whilst the price before named would bring it to 2,984 francs.

In order to answer Mr. Jenkins's last question, I add to this note some printed documents designated as follows:

Instructions upon the organization and superintendence of the service of lights and beacons — 1842.

Schedule of charges and detailed estimates relative to the furnishing of the oil of colza — 1848.

Detailed estimates of the annual expenses of the service of light-houses and beacons — 1848.

Rules for the keepers of lights and beacons — 1848.

Two government circulars (20th and 28th November, 1848) upon the new organization of the light-house service.

Instructions for the service of lenticular lights — 1848.

Instructions for the service of beacons — 1848.

5

Circular (20th December, 1848) of the engineer-in-chief, charged with the service of lights.

Light-keepers' book.

Description of the light-houses and beacons lighted on the coasts of France on the 1st January, 1851.

I will add that, since 1845, the construction of lenticular apparatus has made marked progress, and that the use of curved mirrors, to transmit to the horizon the rays which pass above or below the dioptric drum, are abandoned ; these rays are collected by catadioptric prisms. which give them the desired direction by total reflection.

This arrangement produces a notable economy of light, and is eminently favorable to the durability of the apparatus.

I regret to be obliged to give so concise a form to the present note, but the letter of Mr. Jenkins only reached me a few days since, although it was dated the 27th of May, and I am on the eve of my departure for an inspection, which will take me from Paris for nearly three months. Yet I am not willing to defer my reply until my return, for this delay would seem long, and might occasion doubts of my lively desire to reply to the honorable appeal of the Light-house Board of the United States. I hope, moreover, that the accompanying printed documents will furnish all the desirable data in the organization of the light-house service of France ; but should I be mistaken, I pray Mr. Jenkins to designate the points upon which any further elucidation may appear to him to be necessary. I will hasten to reply to such inquiries, happy to furnish a great and illustrious nation with that experience which has been acquired by long practice and conscientious studies.

I pray the Light-house Board of the United States to accept the expression of my sympathy with their labors, and of my respectful devotion.

L. REYNAUD,

Eng. in Chief, Sec. of the Light-house Commission,
charged with the direction of the service.

Paris, *July* 23, 1851.

P. S. I add to the documents enumerated in the present note others which might be considered of some use :

1st. Drawing, giving the arrangement of a chimney, surmounted by its regulator, of sheet iron.

2d. Two drawings of models adopted by the French government for the building of beacons.

3d. Directions on the placing of lanterns and of apparatus for lenticular beacons.

There have been no written instructions for the establishment of light-houses, because this difficult and delicate operation is always directed by one of the conductors attached to the central service.

———

PARIS, *July* 28, 1851.

SIR : I have the honor to acknowledge the receipt of your esteemed letter of the 29th May last, and the valuable documents which accompanied it.

I observe with great pleasure that the United States government favors the introduction of the lens system of lights, and that a regular organization, similar to that which we have in France, is proposed.

Such an administration, composed of such learned and able men, cannot fail to produce, in a short time, great improvements in the illumination of the coast, and greatly to the advantage of commerce and navigation, which is so extensively and actively carried on in the United States.

You will find herewith, sir, a brief account of the different improvements which I have introduced into the lens system since your visit to Paris in 1846, and the number of lenses which I have con-structed since that time.

I hope that the occasion may present itself to enable you to judge, from your own observation, of the importance and value of these changes ; and I shall employ all my efforts to merit the approbation of the Light-house Board, and of yourself.

I have the honor to be, with great respect, your obedient servant,

HENRY LEPAUTE,
Constructor of Lens Apparatus, No. 247 *Rue St. Honoré, Paris.*

To Mr. THORNTON A. JENKINS,
Secretary to the L. H. Board of the United States of America.

Paris, *July* 28, 1851.

Note upon lens apparatus, and upon numerous improvements in the optical and mechanical parts of them, by Henry Lepaute, constructor of lenticular apparatus, No. 247 Rue S^t. Honoré, Paris, France.

The constant and uninterrupted study, by the undersigned, of the construction and management of lens lights, has enabled him to introduce modifications in the mechanical lamps, which have rendered their repair much easier.

Escapement lamps, upon the plan of Henry Lepaute.

As some inconvenience has resulted from the manner in which the escapements were attached to them, their form has been changed, and they have now had a solidity given to them which remedies that defect.

The levers which bind the escapements to their vertical axes were formerly of unequal lengths, from which it frequently resulted that inattentive keepers sometimes substituted one for the other, thus producing a great inequality in the flow of the oil. Now, however, by a new arrangement these inconveniences no longer exist, and it is found in practice that the management of it is greatly facilitated.

Notwithstanding the perfection to which these escapement lamps have reached, and the great attention given to their management, I have, under the direction of M. Degrand, under engineer to the French light-house establishment, constructed a new lamp, which seems to combine all the qualities of simplicity, durability, and facility of being repaired, which ought to be expected in this description of apparatus, to make them as near as possible within the range of the comprehension of keepers and others who are strangers to mechanics.

These new lamps are founded on the system vulgarly called the "moderator." Each lamp is composed of an iron frame, similar to those of the old lamps, supporting a cylindrical reservoir, in which a piston is placed and adjusted, which is charged with regulators of lead in sufficient quantities to elevate it to the summit of the burner; a small steel cylinder, upon which a chain is wound, serves to raise this piston by means of a crank, and in this ascending motion, the oil with which the reservoir is filled above the piston passes beneath; the action of the piston in descending causes the oil then to rise into the burner, and the overflow to fall back upon the same piston.

The extreme simplicity of this lamp, the small number of pieces of which it is composed being very solid, appears well adapted to the service of light-houses, and requires that the keepers should possess but ordinary ability to enable them to manage them.

The undersigned takes the liberty, then, of calling the attention of the Light-house Commission of the United States to this new system.

Numerous improvements of detail have besides been added to the burners and to the means of carrying off the smoke; the principal change consists in a new ventilator, in which a damper or regulator is adjusted, which by these means does not rest any longer upon the glass chimney of the burner, and permits the exterior current of air, which is introduced by the lower part of the ventilator, to carry off the smoke rapidly from the burner; the strong current giving a more brilliant flash, and preventing the action of the exterior currents of air upon the top of the regulator, diminishes strikingly the breakage of the glass chimneys.

In 1849, the undersigned presented to the French industrial exhibition, at Paris, a lens apparatus of the first order, for short eclipses, upon a new plan, giving in the space of two minutes a flash of sixty seconds duration, followed by an eclipse of twenty-five seconds, which produced a brilliant and powerful flash of ten seconds duration, which was succeeded by another eclipse of twenty-four seconds, and successfully during the entire night.

This apparatus received the approval of the Savans, and the author of it received a new gold medal. It had been constructed by order of Capt. Stansbury for the Carysfort light-house, and was shipped to the United States at the close of the exhibition. I am ignorant of what has since become of it, and its value has not yet been paid to me.

In January, 1850, the undersigned conceived the idea of a new combination of catadioptric apparatus for eclipses, capable of producing flashes equal in duration to the eclipses, and arranged in a manner to show each five seconds a brilliant flash of five seconds duration.

The new character of these lights failed, for we could not reach the result which Mr. Stevenson, of Scotland, had obtained when in giving to the lens apparatus, constructed up to that time, a rotation so rapid that rendered the flash too indistinct, and would not permit the navigator to see it for a sufficient length of time.

A lens apparatus was ordered of the undersigned, by Mr. Lewis, for the Sand Key light, and its execution was well advanced, when, in the month of March last, the order was received to suspend the further execution of the work ; not having received any new instructions, this work remains unfinished.

This year the undersigned is about arranging a new mode of constructing annular lenses for eclipse lights ; that modification, which has also received the approbation of the commissioner of lights in France, consists in the new division of the lenses, of which the elements divided vertically will present an angle of four to five degrees, which will greatly increase the duration of the flashes and will shorten equally that of the eclipses.

The French administration is about to order from the undersigned an apparatus of the first order of this description of flashes for every minute, to renew the apparatus of the light of Ailley, near Dieppe, and another of the same order for eclipses of every thirty seconds for the light of the Baleines, near La Rochelle.

In these new apparatus the lenses which had formerly one metre of elevation are increased to 1m.30.

Last year the undersigned commenced the construction of a lens apparatus of the first order for the Brazilian government, to be placed upon Mount St. Paul, near Bahia. The apparatus of the first order of total eclipses for the new light, is of a combination entirely new. The twelve annular lenses, which constituted the dioptric drum, are surmounted by twelve catadioptric pannels, transmitting to the horizon the flash of twelve small annular lenses placed in the interior of the apparatus, which augment and prolong strikingly the flash of the lenses.

Above the dioptric drum are placed four fractions of prisms, which, turning with the lenses, add also their power.

In this apparatus the entire optical part is revolving, and there is no fixed light during the eclipse ; for it seemed to the undersigned that in consequence of the length of the flashes there would be the advantage of rendering the eclipse total, while the flash of the fixed light of revolving apparatus could not in any case be seen so far as the flash of the lenses ; and besides, it seemed better to increase the intensity of that flash and increase the duration, by rendering the eclipse total. That combination is generally preferred by the officers of the navy, to whom the undersigned submitted his plan before putting it into execution.

The result of these modifications is, that the revolving apparatus of the second order would equal the power of the old lens apparatus of eclipse of the first order, and naturally produce a great economy in the acquisition of equality of light.

The difficult construction of these new apparatus renders the diminution of price impossible, and the undersigned refers for their value to the tariff furnished in 1845.   •

HENRY LEPAUTE,
Constructor of lens apparatus,
No. 257 Rue St. Honoré, Paris.

PARIS, *August* 10, 1851.

SIR: I had the honor to receive your letter of the 29th May from Washington, on my return to Paris, and hasten to transmit through your legation, agreeably to your request, numerous documents which will serve as replies to the questions which you have been pleased to address me.

I have the honor to be, sir, your very obedient servant,

T. LETOURNEAU.

Lieut. T. A. JENKINS, U. S. N.,
     Secretary to Light-house Board, Washington.

—

NOTE UPON LENS APPARATUS.

IMPROVEMENTS IN THE OPTICAL AND MECHANICAL PARTS OF THE APPARATUS.

Considerable improvements have been introduced, or are about to be, in the combination and composition of lens lights. I will treat of them here in a general way, accompanied by a brief description.

Although the most of these improvements are due to my researches, allow me, for the moment, to forget the duty which under all other circumstances modesty would impose upon me, and attach to each of them the name of the author.

Holophotal apparatus, by Mr. Stevenson.

This new combination converts the catadioptric zones into annular zones, thus serving to extend the lenses of the centre the total height of the optical part of these lenses which is contained between No. 6

of the zones of the upper and the No. 10 of the lower part. That combination is only applicable to revolving lights, called flashing, every minute or every half minute in all the different orders.

In taking for a basis of prices those indicated in our notices of these apparatus, in so far as relates to the frame and the optical portions, it will be ten per cent. more expensive.

New combination in the lenses of the drum in the apparatus of short eclipses, by T. Letourneau.

This arrangement consists in replacing the lenses of vertical elements upon an exterior frame, making its revolution around a drum of a fixed light by a system of fixed light lenses and annular intercalary ones between them; the light will remain in appearance otherwise the same; that new made presents a useful simplicity, and a striking economy in the cost of the optical parts.

Combinations common to apparatus for revolving lights, either for flashes for every minute, half minute, or for short eclipses, by T. Letourneau.

At the same time that Mr. Stevenson made his researches upon his new holophotal lenses, the thought of increasing the intensity of the flashes occupied my attention also. I have constructed an apparatus in which I have applied before these catadioptric zones, the same lenses of cylindrical elements which I reduced to the apparatus for short eclipses, (the emergence of the light which is produced by these catadioptric prisms is, as is well known, the same as in the fixed light drum;) thus placing in the prolongation above and below the central annular lenses, the lenses of prisms converging horizontally. I have obtained in the same point the coincidence of the flashes of the crown lenses.

Thus composed, this apparatus is found to be materially increased in value, proportionally to the number of supplementary lenses of which it is composed; but the placing, so far as relates to the frame, does not exceed the small increase of 1,500 francs, and the solidity of the system presents better assurance of durability and of regularity than those apparatus of the same kind ordinarily.

Annular lenses to prolong the flashes in revolging lights, (Administration Française.)

The lenses are divided into two parts, and by their centre of figure upon the height; placed in a manner to form an angle of some degrees, the duration of the flashes increases proportionally to that inclination. but at the expense of its intensity. That combination of lenses gives place to an increase of ten per cent. upon the cost of the frame and lenses.

The same lenses for the duration of the flashes, by T. Letourneau.

Encountering considerable difficulty in the construction of the existing frames (the preceding system, however much it may be necessary to regulate the uprights to the demands of the bed of the frame,) in place of dividing horizontally the lenses, I divide them vertically; the result is absolutely the same. This combination of lenses gives an increase of ten per cent. for each lens.

Revolving lights with alternating movement, by W. Wilkins and T. Letourneau.

In a large number of lights, a small angle of the horizon is only required to be illuminated; and if we can, in a fixed light, propor tion the number of lenses of the drum to the field of the illumination required; it is not the same in revolving lights, as the drum ought to be complete. Having been solicited by many orders which have been addressed to us, treating of this subject, we have searched for the means by which to realize—

1st. The advantage sufficiently striking in economy which would result from the suppression of several lenses.

2d. That which is not less important in utility by the means of reflectors, of the light directed from the land side. An alternating movable machine, whatever the special disposition of the frame, insures to us a movement very regular of going and coming, which circumscribes the march of the revolving lenses in an eighth or sixteenth of revolution.

Glass reflectors, by T. Letourneau.

For the old metallic reflectors, the inconveniences of which are appreciated, I substitute reflectors of mirror glass (silvered.) These new reflectors are placed at a distance from the lamp, which varies with the horizontal amplitude, following the number of suppressed pannels, and of the distance of the focus from 0.307 to the back of the reflector against the interior side of the lantern. Let us conceive that the position the most favorable to place the reflectors is the farthest from the focus, since the divergence is least sensible.

Improvements of detail in construction of frames, table for service, changing the lamps, &c., by T. Letourneau.

One of the greatest difficulties which presents itself in mounting a first order lens apparatus in its tower is the placing the table of service, which up to this time is made of a single piece. I have divided them in the lights which I have fitted up in four pieces, assembled by screws and nuts; an opening arranged under the stair allows the keeper of the watch to place alone and instantly the extra lamp. These arrangements increase the price of the frame three per cent.

OLD SYSTEMS.

Floating lights, fixed and revolving, with lens apparatus, by T. Letourneau.

The floating lights at present in use are composed of a greater or less number of metallic parabolic reflectors, adjusted around the masts of the vessels upon which they are placed. They are divided into fixed lights, and varied by flashes. Besides the inconvenience due to the multiplication of lamps, there is not a sufficient quantity of light produced by this apparatus.

NEW SYSTEMS.

The new floating lights are, in the first place, of four refracting apparatus for fixed lights, embracing an angle of 212 degrees. Secondly, of spherical reflectors, making for the optical part the same focus which it should have with the horizontal amplitude of 148 degrees. These apparatus represent very nearly a weight equal to that of eight reflectors which they replace, and an illumination each by a lamp consuming 0.045 grammes per hour.

They are placed in suspension, equidistant from each other, and in whatever manner, always directed to the horizon, and whatever point occupied by the observer, a light greatly increased.

The prices of these apparatus are fixed for the set of four optical parts, reflectors and lamps, 4,800 francs. For the lantern 3,000 francs. In general the old lanterns may be used for new apparatus.

Revolving machinery for floating lights, by T. Letourneau.

As much from communicated as personal observation, I have been convinced of the inconveniences arising from the placing of rollers between plane parallel surfaces, as very little time suffices for them to make a succession of hachures or indentations sufficiently marked to change the regularity of the revolutions, and in the hands of incompetent keepers are operations of difficulty.

To obviate these inconveniences, this is the way I have proceeded. Preserving the same number of rollers, I caused them to be placed between two conical surfaces having a common centre; in that way I have obtained the regular development of movement following the generation of the surface of revolution.

Lens apparatus for lighting sea steamers from their wheel-houses and bows, by T . Letourneau.

I find myself conducted in descending the ladder of applications of optics, to acquaint you with the last modification, that of lights to prevent collision at sea.

The result of detailed experimental researches made by some of my friends among the officers of the French navy, has been to engage me to construct, according to the ordinance regulating it, small apparatus for colored lights to produce a light sufficiently intense to be distinguished (if a white light) ten miles in good weather. In consequence of the success with which I have met in these lamps, I find myself within the last two years charged with the supplying of all that are required for the navy and for the commercial marine, both of which are subjected in the ordinance to the same general rules.

I do not know if the United States has taken steps in this important matter similar to those of the British Parliament and of the ordinance of Gen. Cavaignac, to insure a uniformity in the manner of distinguishing steamers at night at sea, and to determine at considerable distances the courses steered.

SERVICE AND REPAIR OF THE APPARATUS.

No changes worthy of note, so far as I know, have been introduced into the regulations relating to the lighting service, except such as relate specially to the modification of the apparatus, and which may have been considered worthy of the attention of those to whom they were entrusted with a view to test their applicability.

Lamps.

There are at present two descriptions or systems of lamps, that of Mr. Henry Lepaute and mine; there is also a third one undergoing trial, but which, however highly recommended for its simplicity, is yet suffering from doubts as to its durability and the regularity with which it will perform. I have had for two months one of these new lamps placed on board of a steamer running at night on the river Rhone; but, besides not having received reliable intelligence in relation to it, that experiment would not prove any thing, even if it were satisfactory. The present question is between the two first lamps.

As you are already aware, sir, that while M. Henry Lepaute's lamp succeeded the Carcel lamp, mine has succeeded his. The system of M. Henry Lepaute was announced, at its introduction, as possessing superior qualities of combination, while mine has travelled on its own way noiselessly.

My lamp is the counterpart of Mr. Lepaute's: it is the solidity of it which gives it the most merit, in my opinion.

To bring at once this delicate subject to a close, I would desire that the commission make a trial of both lamps.

Lamp wicks.

For some time past I have been making trial of different tissues for wicks. The object of these researches was to get rid of the operation of trimming the wicks during the night. I have long known how prejudicial the operation is to the service, but I have been compelled to abandon the subject.

Lens apparatus constructed by me for France, England, Scotland, Ireland, Spain, &c., since 1835.

In furnishing the list of apparatus constructed for these different countries, I cannot say that it contains all; by other hands you will receive the complement. Mr. Wilkins, of London, informs me that he will forward to you a list of the lights put up in England, Ireland and Portugal, the great part of the optical parts of which I have supplied. You will find from page 3 to 7 of my notice, the table of apparatus constructed by me before and since 1845, excepting those of England, Ireland and Portugal, which I have sent often without knowing their destination.

Value of the apparatus.

The price of the apparatus has not changed.

Oil for illuminating light-houses.

All the experiments which have been making for a long time upon the different kinds of oils, have given results showing them to be in every respect inferior to that obtained from colza. The rich oleaginous quality of that grain, and the quantity of land prepared for its culture, have rendered it of great importance ; besides which, the oil is now prepared in the most perfect manner. A friend of mine in Normandy, who is familiar with the modes of preparing the land for the culture of this plant, has promised to address me shortly on the subject, and furnish to me in detail an exact account of the process. For the present, I must confine myself to the following hints :

Rich land is preferred for the cultivation of the colza; but, however, it is commonly the custom in Normandy to sow the colza upon the fields from which wheat has been cut. The north of France is its peculiar region.

Markets.

The principal markets for colza oil are : for the north of France, Lille and Coutrai, and for the west, Caen and Rouen. The price is necessarily variable, its average rate is 72 francs the hectoliter for the first quality clarified.* The prices may differ greatly at certain ports for transhipment, depending upon the small or large number of emigrants applying for passage.

MARINE SURVEYOR'S OFFICE,
  Liverpool, November 20, 1851.

SIR : I regret that my absence from Liverpool for a long period, in consequence of ill health, should have caused your letter from the Light-house Board of the United States to remain so long unanswered.

Having only recently returned to my official duties, I now hasten to reply to the question proposed in your communication of May last, which, I trust, will be found satisfactory, and should further information, or explanation of any matter connected with the marine depart-

* A hectoliter is a little over 26 gallons, making the price of the first quality clarified colza oil in France 55½ cents per gallon, while sperm oil ranges in this country from $1 20 to $1 50 per gallon.

ment at this port, be required by the Light-house Board of the United States, both the dock committee and myself will have much pleasure in furnishing it.

1. No change has been made in the mode and materials for light-houses at this port.

2. The best description of mortar for resisting the action of salt water, &c., is that composed of sand, lime, smithy ash ; and the best coating for light towers, is paint made of pure white lead, with sand thrown on it while the paint is in a liquid state.

3. No *Fresnel* lights have been introduced here. Parabolic re-flectors of twenty-one inch diameter are in general use at the light-houses, with the exception of the two Hoylake light-houses, in which parabolas of twenty-nine and twenty-seven inches are used.

Refined olive oil has been introduced and burnt at the various light-houses and light ships, instead of sperm oil. Since the year 1847, the saving in cost price, as compared with sperm oil, is about forty per cent.; and the illuminating powers are fully equal, if not superior, to those of sperm oil.

4. The best reflector in use is the parabolic. The proportion of silver used here is six ounces to the pound avoirdupois of copper. They are made of various prices and qualities, varying from two ounces to six ounces of silver to the pound of copper. Their illumi-nating powers and focal distance of lamp burner from reflector, are tested with those in use by means of the intensity of the light re-flected by them.

5. Argand lamps are used in reflector shore lights.

6. The ventilation in the shore lights is effected by a chimney and turn-cap on the summit of the light-room or lantern ; air valves in window frames and sides of light-room ; and copper tubes over the lamp cylinders to carry off the smoke.

7. The materials of lanterns in the light-houses and sizes of glass vary considerably. In the Crosby light-house, (one of the last erect-ed,) the lantern is of wood, lined with sheet iron ; the window frames of iron ; the panes of glass, of which there are three, are of the di-mensions of six feet by three, glazed with putty and pins. The as-tragals are vertical, and although very strong, obstruct very little light. In the Rock light-house, built after the plan of the Eddy-stone light-house, the window frames and roof of lantern are of copper ; there are sixteen squares of glass in the round, four ditto

in depth; the squares are two feet five inches by two feet; the lantern is fifteen feet six inches in diameter; the astragals are vertical and horizontal.

8. The burners of the lamps are tipped with silver to prevent their too rapid decay from the action of the flame.

9. The best parabolic reflectors, after being twenty years in use with good care and attention, are not perceptibly deteriorated. The inferior quality are not used here.

10. A set of the best lamps, with silver-tipped burners, with good care and attention, will last twenty or thirty years, with occasional repairs and the renewal of the burners, which last on an average three or four years.

11. The repairs of illuminating apparatus are made under the direction and superintendence of the dock and marine surveyors.

12. The lamps and reflectors in the light-ships are not mounted on gimballs; but parabolic reflectors of 21 inches diameter, similar to those in the light-houses, are used in the light ships, and lamps with a flat wick two inches wide; the oil cistern is placed at the back of the reflector, and the cistern is ingeniously subdivided by partitions perforated with holes, to prevent the flooding of the oil by the ship's motion. The introduction of these improved light-ships, lanterns, and reflectors, by which a light is produced equal to that shown from the light-houses, is due to Mr. Hartley, dock surveyor to the Liverpool dock trustees.

13. The greatest improvement made in the light-ship's moorings, is the introduction of wrought-iron welded studs in the chains, instead of the common cast-iron studs. These improved chains are used in the thirty fathom lengths next the bows of the light-vessels, and the chains have been found to break much less frequently than they did prior to their introduction. The remaining portion of the light-ship's moorings are of the ordinary stud-chain.

The buoy moorings vary in size, according to the dimensions of the buoys; the chains are made without any studs; the sinkers are of the shape and size shown in the accompanying drawing.

14. The fog signals in use at the light-houses and light-ships consist of bells and gongs.

15. The material I should recommend for buoys, floating beacons, and light-vessels, is *iron*. The colors, shapes, distinguishing marks, &c., are shown on the chart of Liverpool bay, of which I send you a copy.

16. The buoys and light-ships are placed and moored in their positions by cross-marks and sextant angles : and a register is kept in the marine surveyor's office of every circumstance connected with them, according to form No. 1, sent herewith.

The light-ships' moorings are hove up and examined as far down as possible, on every favorable occasion : they are entirely hove up, and the worn parts renewed annually. Duplicate light-vessels and buoys are kept ready for instant use. A buoy-tender manned by the shore party of the light-ships' crews, executes the buoy service, heaving up the moorings, &c.

The number of officers and men employed on board the light-ships, will be found in the light-ships' rules sent herewith.

The masters of the light-ships are relieved every month : the seamen serve two months on board, and one month on shore.

For relieving shipwrecks, &c., the dock committee have established five life-boat stations in Liverpool bay, the rules and regulations for which I send herewith ; also a copy of my report to the shipwreck committee, which contains all the information you require under this head.

The annual expense of this branch amounts on an average to £1,900.

Every branch of the marine department is inspected monthly, either by the marine surveyor or his assistant.

17. The appointment of light-keepers, &c., are made by the dock committee.

The instructions to the light-keepers are enclosed herewith. numbered from 7 to 13.

18. Iron-rod electrical conductors are employed.

19. The oils are tested by their specific gravity. burning, and the intensity of the light produced.

The chain moorings, both for buoys and light-ships, are tested at the corporation testing machine.

20. The system on which lights are masked at this port, is shown on the accompanying chart of Liverpool bay ; the object being to point out to the mariner, by the opening and shutting of the lights, when he is to alter his course in the channel, or at the elbows of banks, &c. It also indicates the position of the Northwest and Crosby light-vessels, in the event of those vessels parting their moorings, and being absent from their stations. It may be effected by means of blinkers placed on the windows of old light-houses, and, in new ones, by the construction of the window frames.

21. The modes of distinguishing lights are *fixed, revolving,* and *flashing;* the colors are *red* and *white.*

The light-vessels show *one, two,* or *three* lights by night; and a large *ball* at the mainmast head by day : they are also distinguished by the color of their paint.

The light-towers are painted *white, red and white,* or *black and white,* according to locality and background ; the sea-marks are painted either *black,* or *white,* or *striped,* according to locality and background.

In renewing our correspondence, permit me to express to you the high sense I entertain of the favorable opinion and disinterested approbation which you were pleased to bestow on the marine establishment at this port; and allow me at the same time to congratulate you on the adoption, by the United States government, of the valuable recommendations and suggestions contained in the very able report of yourself and Lieutenant Bache, in 1846 ; and wishing you every success in the highly important and arduous duties that lie before you, I beg to subscribe myself,

With sincere esteem and respect, your obedient servant,

WM. LORD.

Lieut. THORNTON A. JENKINS,
Sec'y U. S. Light-house Board, Washington.

MARINE SURVEYOR'S OFFICE,
Liverpool, October 16, 1843.

SIR : In compliance with the instructions of the sub-committee on shipwrecks, that I should report on the life-boat establishment of this port, and the extent and nature of the banks in Liverpool bay, I have to submit the accompanying statement :

There are nine life-boats stationed as follows, viz :

Liverpool	2 boats,	1 master,	and	10	men.
Magazines	2 "	1 "		10	"
Hoylake	2 "	1 "		12	"
Point of Air	2 "	1 "		10	"
Formby	1 "	1 "		12	"

Nearly all the boats have been built since 1839. They pull double-banked, are rigged with two sprit-sails and a jib, are of large size, possess great strength, and are constructed on the most approved

6

principles, with air-tight casks inside, and a broad band of cork running round the whole length of the boat, above the water-line, to resist violent shocks and give increased buoyancy, enabling the boat to float, although loaded with a considerable number of persons and filled with water ; as many as fifty individuals having on one occasion been rescued from a wreck at one trip, making, with the boat's crew of eleven, sixty-one persons in the boat at one time.

The boats are kept on carriages in the boat-houses near the shore, and horses are provided to enable them to proceed to the most advantageous spot for launching. A gun is placed at the station to summon the crew, as also distance-flags at each light-house, light-ship and telegraph station, for the same purpose ; the arrangements in these respects being such, that in many instances the life-boat has been manned, launched, and on her way to the wreck in seventeen or eighteen minutes from the time of the distress-signal being seen.

The masters and crews of the Hoylake, Magazines and Formby boats are composed of picked fishermen, intimately acquainted with the banks, swashways, tides and currents in Liverpool bay. They reside in the immediate vicinity of their respective boat-houses.

The Liverpool boat's crew consists of experienced boatmen residing in the town.

The Point of Air boat's crew consists of two experienced Hoylake fishermen as master and mate of the boat. These men have been engaged for the last four years, at an increased annual salary, expressly for the purpose of organizing this boat's crew, and the rest of the crew are selected from the best and most expert men that can be found in the neighborhood.

The whole of the crews are kept in constant and permanent pay. They are regularly mustered and exercised once a month, and no expense has been spared in rendering the boats, their equipments, and crews, as perfect as possible.

The Hoylake boat is under the active and vigilant superintendence of Mr. Sherwood, revenue surveyor at Hoylake; the Point of Air boat under that of Mr. Dawson, of Grovant, and the Formby boat under the direction of the keeper of the Crosby light-house.

The banks and dangers in Liverpool bay may all be comprised within a triangle formed by the western patch of West Hoyle, the Rock light-house, and Mad wharf—the former (West Hoyle) distant from Liverpool nineteen nautical miles, the latter (Mad wharf) twelve nautical miles; the cross distance between West Hoyle and Mad wharf being sixteen miles.

The principal banks are West Hoyle, East Hoyle, Great and Little Burbo, the Jordan flats, Burbo flats, Mad wharf, Mockbeggar wharf, Formby bank, and Taylor's bank. They are all remote from Liverpool, and many of them several miles distant from the nearest life-boat station—many patches of all the banks dry, and some of them to a considerable height above the low-water level; but in heavy, on-shore gales of wind, owing to the shallowness of the water, a continuous line of heavy breakers extends far to seaward on the weather side of them, rendering it extremely difficult, and at times perfectly impracticable, however advantageously the boats may be placed, and however near they may attain the position, to penetrate to a wreck so situated, without the certain and inevitable destruction of the life-boat and her crew. This has been strongly exemplified in the cases of the Athabasen and the Despatch, lost on West Hoyle. In both instances, the life-boats reached the scene of the disaster, although four miles distant from Point of Air, and ten from Hoylake stations, but were obliged to abandon the crews to their fate, after repeated attempts to penetrate the line of breakers extending outside of them. Numerous instances of a similar nature might be adduced, occurring to vessels lost on the sand banks on the east coast of England; one, in particular, recurs forcibly to my memory—that of the "Ogle Castle" Indiaman, lost on the Goodwin sands. In this case, the Deal boats were so near as distinctly to see the successive surges sweep away the unfortunate crew from the deck and rigging, without being able to render the slightest assistance.

The remoteness of the banks from the land, and the long line of shoal water extending from them, renders the use of all projectiles, such as mortars, rockets, &c., inapplicable in Liverpool bay, as it very seldom happens that a wreck can be approached sufficiently near to render them available.

In face, however, of all the natural difficulties which Liverpool bay presents under these circumstances, and with the immense amount of shipping which enters and quits the port, it must be a gratifying fact for the committee to know, that for the last four and a half years only one call (that of the "Despatch" sloop, lost on West Hoyle) has occurred, in which the life-boats could not render assistance, owing to the circumstances above shown; whilst in the same period no less than seventy-four vessels have been assisted, and four hundred and fifteen lives preserved by their life-boat establishment, owing, no doubt, chiefly to the judicious position of the life-boat stations, by which one or other of the boats has always been able to reach the scene of the wreck; and lastly, to the perfect confidence the crews feel in their boats, and the praiseworthy exertions they have invariably shown in the performance of their duties.

To facilitate the operations of the life-boats, an arrangement is in existence with the steam-tug company, by which one of their steamers is to proceed out immediately the signal of distress is seen flying, taking in tow the first life-boat that reaches her, whether their own or one belonging to the Dock Trust, or both, if the weather will permit of it.

In order to point out and identify the exact spot to which the steamer or life-boats are to proceed, the whole of Liverpool bay and the coast, thence to Holyhead, has been divided into squares, numbered consecutively on the chart, each light-house, light-ship, and telegraph station, &c.. being provided with such chart, and the keeper directed to report, by signal, the number of the square in which any wreck may occur.

I am, sir, your obedient servant,

WM. LORD.

To the Chairman of the Sub-Committee on Shipwrecks.

Letter from Thomas Stevenson, F.R.S.E., F.R.S.S.A.

EDINBURGH, *August* 7, 1851.

SIR: In the absence of my brother, I beg to acknowledge the receipt of your letter of the 27th May.

Having carefully perused the list of queries embraced in your communication, I beg leave to direct your attention to my brother's account of the Skerryvore light-house, and notes on light-house illumination, published in 1848, and to his more recent publication entitled "A Rudimentary Treatise on Light-houses," &c., published by Mr. Weale, of London, in 1850. I think you will find answers, so far as they are known, to your queries in both of these books, but especially in the "Rudimentary Treatise." I may state the following, however, as they may not be so explicitly given in either treatise.

1. I am not aware that any of the slide lamps have ever been worn out.

2. Nothing has occurred to shake confidence in the concentric burner of Fresnel.

3. The same kind of oil is used in the northern light-houses, both in summer and winter.

4. The electric conductors, which are of copper, are three-fourths of an inch in diameter, and tipped with platinum or palladium.

5. There is no difference in the number of light keepers for lens lights and catoptric lights.

6. No light keeper is allowed to leave the light-room a moment before the arrival of his successor.

7. None of the northern light-houses have their characteristic distinctions altered.

I send you, with this, three copies of a pamphlet just published, which contains the most *recent* improvements of which I am aware. These improvements constitute what I have called the Holophotal system of illumination, which, by transmitting *all* the divergent rays of a flame without unnecessary reflection or refraction in one parallel beam, thus gives out the maximum effect of the lamp. You will find a reference to this system at p. 142, part I, of my brother's treatise.

Although only very recently proposed, we have successfully introduced it at five different places, and another light is just about being got ready.

I remain, sir, your most obedient servant,

THOMAS STEVENSON.

Lieut. THORNTON A. JENKINS, U. S. N.,
Secretary to Light-house Board, Washington, D.C.

Northern Lights.

LENS COMMITTEE, *February* 23, 1833.

Present: J. A. Maconochie, esq., James L. Amy, esq., Andrew Murray, esq., Archibald Bell, esq.

The convener laid before the meeting the following communication from Sir David Brewster, in regard to the comparative merits of lens and reflector lights.

"I take the liberty of addressing you as convener of the Light-house Committee, on the comparative value and economy of lenses and reflectors, and of suggesting some additional experiments for exhibiting the superiority of the former. I need not mention to you, that the most distinguished philosophers in the institute, and the most celebrated engineers and naval officers in France, have, on the authority of direct experiments, decided in favor of lenses, both as to effect and economy.

"The Academy of Science at St. Petersburgh have done the same more than five years ago, and Sir John Herschel has lately published the following testimony in their favor. 'The light-house,' says he, 'with the capital improvements which the lenses of Brewster and Fresnel, and the elegant lamp of Lieut. Drummond, have conferred and promise yet to confer, by their wonderful powers, the one of producing the most intense light yet known, the others of conveying it undispersed to great distances.' Disc. on Nat. Phil., p. 56.

"The superiority of lenses, however, is no longer a matter of opinion, for it was proved, in the presence of the committee, that a single lens was equal to at least nine reflectors.

"This lens, too, was only twenty-nine inches in diameter, and was of green crown glass. Had it been made of good flint glass, and been thirty-six inches in diameter, it would have been equal to at least fourteen reflectors, four being added for the increase of area, which is a matter of simple calculation, and one for the loss of light in the green glass. I shall not avail myself, however, of this undoubted fact, but shall take the result which was actually seen by the committee, namely, that one lens equals nine reflectors, and I shall apply it to the case of a revolving light.

"The revolving light with lenses will consist of two lenses placed opposite to each other, and illuminated by a single lamp between them. The revolving light with reflectors will consist of eighteen Argand burners, nine reflectors being substitutes for each lens.

"It being admitted that these two pieces of apparatus will give the same light, let us consider their comparative advantages:

"1st. The lens apparatus will be decidedly the cheapest in its first cost, and the lenses will never require to be renewed.

"2d. The lens apparatus will not require one-third of the labor in cleaning and arranging them daily for use.

"3d. The lens apparatus will not require so strong and powerful a piece of machinery to move it, from its inferior weight and its greater compactness.

"4th. The lens apparatus may be placed in a much smaller light-room, the eighteen reflectors requiring a very large space, and economy might thus be introduced in the erection of future light-houses.

"5th. The eighteen Argand burners will decidedly consume more oil than the simple compound burner used for the lenses; hence it follows that the lens apparatus is in every respect better and more economical than the reflector apparatus.

"But in the above comparison I have omitted entirely the addition of the lateral lenses and mirrors which I proposed in 1811, and which were used in the French light-houses. These lenses and mirrors will enable us to take as much light from the compound burner (which otherwise goes to waste) as will be equal to four reflectors at the very least; so that without the slightest additional expense, excepting the original cost of these small lenses, &c., we can give the lens apparatus a great superiority over the reflector apparatus.

"Let us now compare the lenses and reflectors in reference to the introduction of extraordinary illumination, which in fogs and in seasons of danger may be required.

"The means which may be thus resorted to are these: 1st, the Drummond light; 2d, the blue and signal lights; and 3d, any extra lamps which may be at the time in the light-house.

"In the lens apparatus the Drummond light can be introduced instantly, by merely putting the lime-ball in the place of the burner. In the reflector apparatus this is impossible. To produce the same effect the eighteen reflectors would require to be each fitted up with a Drummond light, the expense of which would be enormous, and for ordinary and extraordinary purposes the light-house would require in all thirty-six reflectors, to be as effective as the lens apparatus.

"In introducing the blue and red lights for occasional purposes, we have only to burn them on an iron plate or dish, placed either beside the burner or occupying its position. From the power of the red light to penetrate fogs, I consider it as an invaluable resource in light-houses. In the experiment which the commissioners witnessed, its brilliancy was fully equal to that of the lens apparatus. I need scarcely add that these signal lights cannot be introduced into reflectors.

"If the light-house keeper is provided neither with the Drummond light nor the blue and red signal fires, he can at present do nothing to add a ray to his reflectors, even if he knew that in a dark and stormy night human life is exposed to danger. With the lens apparatus, on the contrary, he can surround the main burner with all the spare Argand burners in his possession, and thus convey a great quantity of additional light into the refracted beam.

"Hence it follows that the lens apparatus is far more intense than a reflector apparatus of the same size ; that with the same intensity of light it consumes much less oil : that in reference to original cost, repairs and renewals, it is more economical; that it requires a less expensive light-room, and demands much less time and trouble from the keeper, while it possesses the property (which reflectors never can have) of admitting every variety of resource in cases which demand extraordinary illumination.

"I have no scruple in stating that the introduction of the lens apparatus will occasion an annual saving in the expenditure of the board: but if this were not the case, and if much more oil were consumed by the use of superior lights, it would still be the duty of the commissioners to adopt it. No mining company uses the old steam engines of Savary and Newcomens in order to save the additional expense of purchasing one of James Watt's. No astronomer continues to use the old refracting telescope because an achromatic telescope is a dearer instrument. In these cases scientific and pecuniary interests are alone concerned : but in the choice of lights, human life is involved, as well as national property, and a higher responsibility is therefore attached to their management. I trust, therefore, that the commissioners, now that they have witnessed the superiority of lenses, will not allow any considerations of economy (even if they did exist) to prevent them from introducing a system of illumination already established in other countries, and calculated to promote the just objects for which their board was instituted.

"In order to prove more fully the value of lens illumination, I beg to suggest the following experiments and calculations :

"1. Try the Drummond light in the focus of the lens, and compare it with the same light in the focus of a reflector.

"2. Try the blue and red lights in the focus of the lens.

"3. Compare a single reflector with the same Argand burner in the focus of the lens.

"4. Compare the quantity of oil consumed by one compound burner with that consumed by eighteen Argand burners of the common size.

"5. Compare the expense of a light-house erected for a revolving lens apparatus, with that of one erected for eighteen reflectors.

"6. Compare the expense of two lenses (£25 each in Paris) with that of eighteen reflectors, adding the expense of their frames and of the machinery for moving them.

"Before concluding these hurried observations, permit me to add, that the use of distinguishing lights is considered by every person as highly desirable, and that it is only with the lens apparatus that they can successfully be introduced. I would beg leave, also, to repeat what I urged on a former occasion, that coal gas should be employed in every light-house where there is accommodation for its proper manufacture. An immense saving would thus be effected, and a more perfect system of illumination obtained.

"ALLERLY BY MELROSE, *February* 16, 1833.

"J. A. MACONOCHIE, Esq., &c., &c."

And the meeting being of opinion that the information contained in the above communication is of the greatest importance, in the view of introducing the apparatus into the light-houses in progress, direct the report of the engineer on the experiments already made, and the above communication, to be immediately reported to the Bell Rock Committee, to ascertain whether it could not be done.

*   *   *   *   *   *   *   *

The difference between the original cost of the reflecting and refracting apparatus is of course more or less in proportion to the number of reflectors or lenses employed. The expense at Paris of the apparatus for the Tour de Cordouan, where there are nine vertical lenses, besides a system of fixed mirrors, was stated by M. Fresnel to be about £1,500; but in estimating a similar apparatus, if again required, he made the amount 28,262 francs, or £1,177 11s. 8d. The cost of the reflecting and revolving apparatus, with twenty-four reflectors, is £1,387. "With regard to the annual expense and other conditions of the lens system, the engineer is not prepared to give an opinion; but after a sufficient number of trials have been made with the very complete apparatus provided by the board for investigating this important subject, he will be enabled to report satisfactorily upon the several objects to which his attention has been directed by the minute of the 23d February. In the meantime, to meet the urgent desire of the Lord Provost of Edinburgh, the above estimate has been prepared from his notes, made with M. Fresnel when at Paris, and from his personal observations of the Tour de Cordouan."

And on motion of the Lord Provost, it was resolved to recommend to a general meeting, to be called for the special purpose, on Monday next, at 12 o'clock, to have the light at Inchkeith immediately adapted to the lens apparatus.

The following communication was since received by Mr. Maconochie from Sir David Brewster :

ALLERLY BY MELROSE,
March 29, 1833.

MY DEAR SIR : As the experiments made on the Calton hill afford sufficient data for guiding the opinion of the board on the relative value and economy of lenses and reflectors, I beg you will submit to the commissioners, at their first meeting, the following statement respecting some important questions to which they will no doubt speedily direct their attention :

1. In establishing a new light-house, should the method of illumination by lenses be adopted ?

2. Would it be wise to dismantle every light-house in Scotland, and substitute lenses for reflectors ?

3. Should gas be introduced in place of oil ?

4. Should means be provided in every light-house for the occasional exhibition of powerful lights ?

5. Should a new system of distinguishing lights be matured and adopted ?

To all these questions I trust the commissioners will agree with me in giving an affirmative answer. That the committee of the Royal Society will concur in the statement I am about to make, I cannot for a moment doubt ; because scientific questions can admit of only one solution. But if, like other prophets, our opinions shall have no weight in our own country, I shall, for the satisfaction of the board, be able to support them by the concurring testimonies of the most distinguished professors in Cambridge and Oxford, and by the practical judgment of the most eminent scientific engineers of the present day.

1. In order to enable me to answer the first of the above questions relative to the introduction of lenses into new light-houses, I have obtained from M. Fresnel, of Paris, a detailed estimate of the expense of fitting up a light-house with the lens apparatus, including all the necessary machinery and utensils, and on the scale which is in use at the magnificent establishment of Cordouan.

The optical part consists of nine lenses, thirty inches in diameter; nine smaller lenses, with their reflectors, for widening the main beam of light; and another piece of apparatus for collecting the light that falls below the lenses. The expense of this part of the apparatus is 16,500 francs, or £687 10s.

The mechanical part consists of all the frame work and revolving apparatus, with three Carcel lamps. The expense of this part, including the smaller utensils, is 9,500 francs, or £395˙16s. 8d.—making the total amount 26,000 francs, or £1,083 6s. 8d.

Now, the expense of a reflecting apparatus with twenty-four reflectors is £1,387, making a saving of £303 13s. 6d. in favor of the lenses, or of £413, reckoning £110 as the value of the plate glass in the lantern for lenses. This saving will be increased to £513, because £100 may be saved by substituting an invention of Mr. Oldham's for the Carcel lamps, or by introducing gas.

If these twenty-four reflectors are arranged in groups of six, then the brightest light which at any one time reaches the eye is that of six reflectors, which is repeated four times in each revolution; whereas in the lens apparatus we have a light equal to nine reflectors, repeated eight times during each revolution, besides the additional light of the eight smaller lenses, and that of the other piece of apparatus. Hence it is demonstrable that the lens apparatus is not only £413, or eventually £513 cheaper than the reflector apparatus, but gives a more intense and penetrating light.

But, independent of these enormous advantages, the lens apparatus is perennial, while the other is perishable, and requires to be renewed.

2. In answering the second question, respecting the propriety of dismantling every light-house in Scotland, and substituting lenses for reflectors, I shall treat the subject commercially, though I might at once solve the question by affirming, that if the lenses afford, as they have been proved to do, a more brilliant and penetrating light, they ought to be instantly adopted from motives of humanity, even if the board were to sustain a heavy loss. It is fortunate, however, that motives of economy and humanity are, in this case, combined in favor of the change. In the following calculations I shall take a fixed light, such as that of the Isle of May, and a revolving light, similar to that of the Bell Rock, and I shall suppose that the apparatus at both these stations is perfectly new.

The Isle of May light-house has twenty-seven reflectors. I propose to substitute, in place of them, a lens apparatus which will cost £1,000, the sum which is saved by there being no revolving machinery being sufficient to purchase the additional lenses. Now, the twenty-seven reflectors at present used will produce £567, and I estimate the value of the lamps, &c., at £50; so that the produce of the present apparatus will, at the very least, be £617, which, taken from £1,000, leaves £383 as the expense to be incurred in establishing the new system of light, which will be at least six or eight times more intense than those now in use. This outlay, however, is nothing when we come to consider the annual saving.

The consumption of oil per annum at the Isle of May is 1,080 gallons, which, at six shillings per gallon, amounts to £324. But the lens apparatus requiring only one lamp, which consumes oil equal to fourteen Argand burners, (according to Mr. Stevenson,) will burn only five hundred and sixty gallons of oil, which amounts to £168. Hence there will be an annual saving in oil alone of £156, besides £10 more for glass chimneys and other articles. This annual saving of £166 will surely justify the commissioners in making an outlay at the Isle of May of £383.

If we now take a revolving light of 24 reflectors, like that of the Bell Rock, the advantage of changing it for lenses will be equally clear. The expense of the largest and most complete lens apparatus for a revolving light, like that of Cordouan, as already in detail, is £1,083. The value of the silver in the 24 reflectors will be £504, which, with £45 for lamps, &c., makes £594, leaving an outlay of only £434 in order to give the Bell Rock light-house a scientific and an efficient system of lights, which will be seen at a much greater distance than the present ones, and through a hazy atmosphere, which will completely obstruct the lights now in use.

But this change will also produce a great annual saving. Twenty-four reflectors consume 960 gallons of oil, which will cost £288; whereas the lens lamp will consume only 560 gallons, which will cost £168—so that there will be an annual saving in oil alone of £120, or £130 including a saving in glasses and other articles. In reference to the economy of light-houses furnished with lenses, I may state to the board, on the authority of M. Fresnel, the highly important and startling fact that the annual expense of the great light-house of Cordouan, including light-house-men's wages, oil, wicks, &c., is £395 6s.,* while the Bell Rock light-house costs the country £861 annually, on an average of four years, and the smaller one, at Tarbetness, £555 annually, on an average of three years.

3. I come now to the question of the substitution of gas for oil. In January, 1826, more than seven years ago, I urged the commis-

---

* Less than $1,800 per annum for Corduan light, fitted with a first-order lens.

sioners to make this great and obvious improvement—an improvement as valuable in point of economy as it is in reference to brilliancy of illumination. Experience, on a great scale, has already given its decision on this subject. Austria has, many years ago, lighted up with coal gas the great light-house of San Salvore, on the coast of Istria, and the effect of the gas far surpasses that from oil lamps. The details of the relative expense of gas and oil have been published by Professor Aldine; but I shall trouble the board only with the grand result:

The annual expense of oil lights was................ 1,861 florins.
The annual expense of gas is...................... 932 ··

Making a saving of........................ 929 ··

Exactly one-half of that of oil. If we include the interest on the money advanced for the gas apparatus, which was 400 florins, the total expense for gas annually is 1,332 florins, leaving still a saving of 529 florins, or nearly one-third of the expense of illumination by oil.

I may add, also, that two small light-houses at Dantzic are fitted up with gas. One of them consists of a parabolic reflector 22 inches in diameter, and the other one of 17 inches in diameter. Here, as in Istria, the introduction of gas has been found to be a measure of great economy, and the additional brilliancy of the lights at Dantzic was so great that the inhabitants of Hela, where the gas was first lighted, attributed the brilliant effect to a great fire in the city.

4. Seven years ago I suggested also the necessity of providing, in every light-house, the means of exhibiting powerful lights in cases of great emergency, and I have described, in my printed paper, the means by which this can be done. The Drummond light is obviously well adapted for this purpose, when placed in the focus of a lens,* but an equally efficacious and much cheaper substitute for occasional purposes has been suggested by Mr. Robinson, namely, the burning

----

* I need not mention to any person acquainted with optics, that in the bungled experiments with the lens and Drummond light, on Thursday, the 21st, and Friday, 22d March, the lens was so misplaced that the brilliant part of the column never reached the eyes of the spectators, who saw only the penumbra; otherwise, the Drummond light, with the lens, would have greatly exceeded that by the reflector. To prove this by occular demonstration, the two lights, at a moderate distance, should be thrown into a room, and their radiative intensities actually measured.

of blue and red lights. I lately recommended the introduction of these lights into the focus of the lens; and the board will recollect the splendid effect which was there produced by the red light—a light especially fitted for penetrating a hazy atmosphere. The use of such lights is impracticable with reflectors, and the facility and effect with which they can be used with lenses, is a new argument, if any were wanted, in favor of the latter.

5. During the late discussions in the House of Commons, Sir Edward Codrington pointed out the disadvantages to seamen of numerous light-houses; but these disadvantages exist only when the lights are not properly distinguished from one another. Hence it becomes a matter of the highest importance, especially when light-houses have become numerous, to establish a well matured system of distinguishing lights. The methods used in France do not afford a sufficient number of palpable distinctions. The use of colored media, or of transparent plates that produce periodical colors, or polarized tints, present us with the best means of supplying this desideratum. Experiments in which I have been long engaged, on the absorption of light by solid fluids and gases, have led me to various results, which are peculiarly applicable to the present purpose. I have in this way succeeded in impressing upon any given light a numerical character which no change can take from it, and which can be recognized by looking at the light through a small and cheap instrument made for the purpose. If the commissioners should desire to witness the effects of this and other methods of distinguishing lights, an apparatus could, with the assistance of Mr. Addie, be easily constructed. But if these new plans should not meet with the approbation of the board, the introduction of the lens illumination will furnish ample means of obtaining distinguishing lights, by a well arranged succession of light and darkness.

Before concluding this letter, permit me to make a few remarks on the proposal to erect a lens apparatus at Inchkeith. As a light-house of the first order is not required at this station, it might, under ordinary circumstances, be sufficient to replace the present very inefficient apparatus by lenses of half the area and half the price (viz: £25 each) of the largest, and to employ only a burner of two wicks, which is equal to three and a half Argand burners.

This apparatus would consume only one-fourth of the oil now used by the seven Argand burners, and would give a light several times more intense. Such an apparatus would cost about £500. But as I presume it to be the object of the board to erect one of the best lens apparatus, as a guide for their future proceedings, I earnestly recommend to them to have the apparatus complete, with all the modern improvements, and especially to light it with coal gas. The expense of the gas apparatus will be paid by its own savings in a few years, and that of the lens apparatus will not greatly exceed £1,000. But, whatever was its cost, it would be creditable to the metropolis of Scotland, and most useful to the navigation of the Frith of Forth, which is frequently beset with fogs, to have at Inchkeith, a lighthouse of the first order in point of magnitude, and of the highest character in point of science.

I have, &c.,

D. BREWSTER.

To J. A. Maconochie, Esq.,
    Convener of the Lens Committee.

Letter from Messrs. Alexander Mitchel & Son.

Belmont, Queenstown, Cork,
July 31, 1851.

Dear Sir : We are in receipt of your favor of 26th May, with the interesting documents enclosed in it, and have much pleasure in replying to your inquiries respecting the screw-pile, being satisfied that the more extensively it is known the more extended will be its application. All the engineers of any standing in these countries have expressed themselves strongly in its favor ; of these we may mention Messrs. Walker and Burgess, Rendal, Brunel, Cubitt, &c., &c., all of whom have adopted it in foundations where the ground was penetrable or covered by the water, and in every case it has proved eminently successful.

The screw-pile light-houses inspected by you and Lieut. Bache, when in this country, are all standing as you left them, in perfect order, no repairs having been necessary, with the exception of paint; since then, some others have been erected by us. In the year 1849, we placed a screw-pile light-house in Dundalk bay for the commissioners of Irish lights; in 1850 we laid the foundation of a screw-pile light-house on the Chapman-sand, off Sheerness, for the Corporation of the Trinity House, London, and we are at present constructing a screw-pile light-house on a shoal in Cork bay, for the Irish Light-house Board : all of which we hold to be a sufficient reply to whatever may have been urged against the system.

The Dundalk light-house stands on nine wrought-iron piles placed seventeen feet in the ground ; it has been some time lighted, and is a great boon to that rising town. The Chapman-sand light-house is also on nine wrought-iron piles placed thirty-nine feet in the ground, and the superstructure is now advancing under the management of the engineers of the Trinity House, Messrs. Walker & Burgess.

The construction of the Cork light-house, the foundation of which is now laid, has, with the exception of the lantern, been entrusted to us by the Irish Light-house Board ; of all which works and others of still greater importance you will shortly have ample details, with drawings, from a small volume at present in the press, some copies of which we shall have much pleasure in forwarding to you ; but probably the most severe test to which these piles have been subjected may be seen in three beacons placed by us on the Kish, Black-water, and Arklow banks, dangerous shoals ten or twelve miles off the coast of Wicklow and Wickford.

Each beacon consists of a massive wrought-iron pile of great length, surmounted by a large ball ; and although depending for support on a single screw at their base, they have all withstood the storms of several winters; but these beacons are not of themselves of much importance, having been placed principally for the purpose of pointing out the sites of future light-houses.

As to our screw-moorings, we have several persons constantly employed in putting them down, in the various bays and harbors of the United Kingdom ; they being now preferred in these countries to all other moorings.

With regard to our light-houses, we have not found it necessary to make any material alteration either in their principle or form, although we have in the materials used, as we now prefer wrought iron and British oak, to the fir timber from the Baltic formerly employed. Wrought-iron piles being in most cases especially necessary by reason of the various descriptions of sea-worm which everywhere infest our coast. Wood, however, is more economical, and may be safely used where the piles are occasionally open to inspection throughout their entire length, as in the case of the Fleetwood and Maplin Sand light-houses.

The slight vibration occasionally observed, so far from detracting from their strength, really renders them more stable, in the same manner and for the same reason that springs have been found useful when applied to heavy wagons; sudden shocks being thus in both cases considerably softened and rendered comparatively harmless; of this the Small's light-house, now standing about sixty years, is a remarkable example, which, together with the objections to cast-iron in sea foundations, and other matters interesting to engineers, you will find alluded to in the pamphlet which accompanies this letter.

Wooden piles when placed in fresh water, being free from the attack of worms, might (should economy be an object) be employed with advantage to support light-houses and beacons on the shoals of lakes and rivers. You say the chief objection urged by those opposed to screw-pile foundations are, too great vibration, excessive torsion, and, in some localities, not sufficient stability to withstand the force of the elements opposed to it. Of vibration we have spoken already, and hold that it is in no way injurious where not so great as to affect the light, and where the structure is placed on a base of sufficient breadth. If by torsion be meant the power employed in screwing in the piles, that can be rendered innoxious by giving to the screws the proper form, and to them and the piles strength sufficient to bear the twist; or, should a rotary vibration of the house be meant, this can be entirely prevented by the application of angle braces between the outer piles as shown in the Belfast and Fleetwood light-houses.

As to giving to such structures power to resist the elements, that must in every case depend on the judicious application of materials of sufficient strength, of not difficult solution, as shown in the success that has attended all our works. Some attempt has also been made in this country to lessen confidence in the screw-pile, by asserting that the Minot's Ledge light-house, destroyed last April, was placed on screw-piles; but this we understand to be a mistake, and would gladly have your authority for its contradiction. The accident was possibly owing to screw-piles *not* having been used in the construction, or too insufficient breadth of base.

We shall, also, at all times be most happy to give you any information in our power on the subject of these works with which we may be connected; and should the presence of one of us tend in any degree to the more extensive adoption of the screw pile or mooring in the United States, our junior is quite prepared to pay your country a visit.

7

It now only remains for us to express our high gratification at your appointment as secretary to this most important commission, and to subscribe ourselves with the highest respect,

Your obedient servants,

A. MITCHELL & SON.

THORNTON A. JENKINS, Esq.

GREAT GEORGE ST., WESTMINSTER,
August 1, 1850.

MY DEAR SIR: I have carefully examined the design for a landing pier at Osborne, in the Isle of Wight, which is, in my judgment, well suited to its purpose, as embracing lightness and elegance of appearance with great strength and firmness of construction, at the same time offering the least possible obstruction to the rise of the tide or stroke of a wave.

These advantages are obtained chiefly by the use of the patent screw piles of wrought iron, which are eminently adapted for the erection of structures of this kind, affording, as they do, a great facility of execution, and in matters of this kind especially, of which I may instance the pier for landing of passengers and goods, driven into the surf or swashing of the sea at Coustoun, in Ireland, in which (as probably it would in this case) the pier formed the staging for its own construction. I am very much in favor of the patent screw, having used it for very large moorings in the river Mersey, and my son, during the last year, built a large railway bridge in the fens of Lincolnshire, founded and supported entirely on screw piles; and I believe they have been extensively used by Mr. Bendel for the construction of piers used in the erection of the great Portland breakwater. Nor can I conceive anything better adapted for a landing pier at Osborne, than a well-framed platform of good timber, supported on piles of wrought iron screwed firmly into the bottom of the sea to the requisite and proper depth ; and were I about to erect a pier, I should certainly adopt the same mode of proceeding.

I am, my dear sir, yours faithfully,

W. CUBETT.

To CHARLES MANBY, Esq.,
Secretary Institution Civil Engineers, London.

NOTES.

During the summer of 1850, Mr. Mitchell laid down a screw-pile foundation for a light-house on the Chapman sands, near the mouth of the Thames. These piles, nine in number and six inches in diameter, were screwed down forty feet, although the bottom is very hard; but this depth was demanded by the contract. The heads of the piles were framed together, and the superstructure is to be completed by the engineer to the Trinity House, Mr. James Walker. Second, Mr. Mitchell has contracted with the Trinity House to put down a similar foundation for another light-house on the Goodwin sands. Third, the son of Mr. Mitchell was employed last December in erecting a screw-pile light-house at the entrance of Cork Harbor, for the Irish Ballast board. Fourth, Mr. Mitchell has been called upon for estimates to erect a screw-pile light-house at Singapore, and also for a viaduct many miles in length, to be constructed over the marshes in the province of Gararat, (India,) terminating in a screw-pile pier for shipping cotton, wool, &c. Fifth, it is proposed to use 2,500 screw piles on the Portland breakwater. These form a viaduct on which the cars loaded with stone from the quarries are made to travel by steam power. The stone is dumped off each car into the water, the piles being thus buried up in the operation. The great expedition and facility obtained by the use of this screw-pile viaduct or railway cannot be over estimated, as the train is loaded at the quarries and then rapidly drawn to the place of deposit, where the trap doors in the bottom of the cars are opened and the whole load discharged in a few minutes. This is one of the greatest works of modern days.

SCREW MOORINGS.

It is highly desirable that these useful moorings should be introduced for securing our buoys—the loss and repairs of which cost now annually $20,000, owing to the insufficient moorings we now use in the shape of iron or stone blocks; a large number of buoys are swept away every year, or else dragged off into deep water or out of true position. The adoption of this mooring is general in Great Britain and the north of Europe.

Report on the Lightning Rods of Light-houses.

SEPTEMBER 25, 1843.

The undersigned have, according to their instructions, met and considered the circumstances under which light-houses are placed as respects lightning, and have arrived at the following conclusions:

That light-houses should be well defended from the top to the bottom;

. That as respects the top, the metal of the lantern and upwards is sufficient to meet every want and satisfy every desire and fear;

That for the rest of the courses down the tower, a copper rod three-fourths of an inch in diameter is quite and more than sufficient:

That at the bottom where the rod enters the earth, it is desirable at its termination to connect it metallically with a sheet of copper three or four feet long by two feet or more wide ; the latter to be buried in the earth so as to give extensive contact with it;

That glass repellers are in every case useless;

That glass thimbles are not needed, but do no harm;

That if the repeller be removed, and the joint on the vane be terminated as the lightning-rods usually are, and then the metal of the lantern be strongly attached to and cemented with the upper end of the copper rod, and the rod continued down the tower to the earth, and the sheet of copper buried in it, such a system will be an effectual and perfectly safe lightning conductor;

That then there need be no rod end rising by the side of and above the lantern;

That the rod may (if required on other accounts) come down on the inside of the building or in a groove in the wall, but should not be unnecessarily removed from observation and inspection;

That all large metallic arrangements in the stone work in other non-metallic parts of the tower of the light-house, such as tying bars, metal flues, &c., should be well connected by copper with the conductor;

That the vicinity of two metallic masses, without contact or metallic communication, is to be avoided;

That as to the South Foreland high light, the lantern, the central stone, and the copper rod proceeding from it to the earth, connected as they now are, form a perfect lightning conductor, even without the rod that is there erected; but,

That it is important casual arrangements should never be depended upon for lightning conductors, but a copper rod be established for the especial purpose; for if the former be trusted to, the carelessness or ignorance of workmen may, at after periods upon occasions of repair or cleansing, cause the necessary metallic connection to be left imperfect or incomplete, and then the arrangement is not merely useless but dangerous.

That as to the Eddystone, it is desirable to connect the system of wrought-iron ties in it with the lightning conductor, by joining the lower part of that iron rod which is nearest to the conductor, with the latter by a copper rod or strap, equivalent to the conductor in sectional area.

That the Dungeness light-house is in a very anomalous condition; to rectify which the two repellers should be removed, and also the representative of the top of a lightning rod attached to the flue; and that then a good copper conductor should be attached to the metal of the lantern, upon the principles already expressed.

M. FARADAY.

Jacob Herbert, Esq., &c., &c.

Extract from minutes of the Board, held on the 26th September, 1843.

Read a report from Professor Faraday on the reference made to himself and Mr. Walker, on the subject of the construction of lightning conductors for light-houses.

Read, also, a letter from Mr. Walker, conveying his approval of all that Mr. Faraday's report contains, and satisfactorily explaining the grounds on which he has not thought it right to sign the report. The board having hereupon considered and approved the suggestions offered in the said report, ordered that Mr. Walker be instructed to cause the light-houses belonging to the corporation to be protected by the means recommended by Mr. Faraday, and in which he expressed his entire concurrence as above said; and that Mr. Faraday be acquainted therewith.

IRISH LIGHTS.

Extract from Senate Document No. 488, 1st session, 29th Congress.

In light-houses, especially fixed lights, where any considerable portion of the circle is to be illuminated, say exceeding one half, (or more than 180°,) refractors should be chosen, as light can with them be more economically used. There is, also, thus a more equal diffusion of the light around. In small harbor lights where an angle of only a few degrees is to be lighted, as in marking a channel, it will be, in most cases, found advantageous to use reflectors.

The general number of keepers to each light-house is two; a principal and an assistant keeper. There are only two light-house stations on the coast of Ireland in which there are three keepers to each; those are on tidal rocks, having shore establishments.

The annexed document (in reply to No. 3) states the number of persons in each light-vessel, and the arrangement for relief.

Buoys are distinguished by differences of color, as black, red, &c., or any of these, with broad paralleled or diagonal belts of different colors, and by affixed beacons of different forms.

Light-vessels are moored by single anchors; (iron mushrooms from one and a half to two tons weight.) The chains are from one and three quarters to two inches in diameter, of iron, with studs or cross-bars. Stone sinkers are generally used for buoy moorings; they weigh from twenty to thirty hundred weight each, according to the size of buoys. The chains are from one inch to one and a half inch in diameter, of iron, without studs. Mushroom anchors were formerly used for mooring buoys, but stone sinkers are found to be as effective (when the buoy alone is to be secured) and are not more than one-eighth the cost.

Weather being favorable, the length of time usually occupied in putting down one of Mitchell's screw moorings for a mooring-buoy is from three to four hours. A strong raft, having a central opening, is best adapted for working the capstan, shears, &c.; but two floats or buoys, platformed over, lashed together, securely moored and well steadied with quarter-lines, form a ready and serviceable substitute.

The ballast board have several of these screws in use at Dublin, and propose using them more extensively, both as moorings and as bases for beacons. Considered as moorings for one or many vessels, although taking the mere weight, they are necessarily more costly than ordinary moorings, yet, bringing into account their much greater power of holding, (in all places suitable to their use,) they would become, perhaps, the most economical that could be adopted.

Time will not now permit me to attempt arranging the foregoing list of replies more methodically or more in detail. The report

which it is expected will shortly be published by order of the select
committee of the House of Commons will contain any further details
that might be wished. GEORGE HALPIN, *C. E.*

DUBLIN, *October* 27, 1845.

Return to an order of the Honorable the House of Commons, dated 14*th*
August, 1846.

A return by the corporation of Trinity House of Deptford Strond,
showing whether they have adopted the recommendation of the
select committee on light-houses in the use of colza, or rapeseed
oil, instead of sperm oil, and what saving of expense has accrued
therefrom; and also whether they have made any and what reduction in the light dues charged by them.

RAPESEED OIL.

The attention of the corporation of Trinity House had been drawn
to rapeseed oil as a substitute for sperm oil, and trials of its qualities,
with the object of introducing it into use at their light-houses and
light-vessels, had been directed to be made prior to the sitting of
the select committee on light-houses, in 1845.

Suitable arrangements for the burning of this oil have since been
made and are continued as opportunities present themselves, and its
use will probably become general at the light-house and light-vessel
stations in England. The quantity which was contracted for and
supplied in the spring of the year 1841 was thirty tuns, being so
much, or nearly so, in diminution of the contract quantity of sperm
oil, the respective contract prices being, for rapeseed oil, 3*s.* 9*d.* per
imperial gallon; spermaceti oil, 6*s.* 4*d.* per imperial gallon. It must
not, however, be supposed that the whole difference in the price per
imperial gallon is actually saved, because the quantity of rapeseed
oil consumed very considerably exceeds that of spermaceti oil, if the
flame in the lamp be maintained, as it should be, at the maximum of
its power.

The time which has elapsed since the use of this oil has been extended has not been sufficient to show conclusively what the excess
of consumption will be in practice; but the present results induce
the expectation that it will be very considerable, and there do not, at

this time, appear any certain grounds for supposing that the difference can be materially reduced.

In the autumn of 1845 the board referred the investigation of the power and qualities of the light produced from the combustion of this description of oil to Professor Faraday, and the following valuable and interesting observations, having reference to the question of consumption, are extracted from his consequent report, dated the 9th day of October of that year, viz: "Having burnt the lamp for many days, I have been much struck by the great steadiness of the rape oil lamps, either considered alone or as compared with the sperm oil lamps. They would burn for twelve or fourteen hours at a time, with little or no alteration of the light, the cottons or lamps not being touched the whole time; whereas the sperm oil lamps would, in the course of four, five, or six hours, give a diminished flame, from the incrustation of the charred part of the cotton retarding the flow of oil. In the rape oil lamps the coal is broken and porous, and serves for wick almost as well as the fresh cotton; but in the sperm oil lamp the coal forms a hard, continuous ring, which seals up the ends of the threads; and this, with the more confined condition of the burner, and the greater distance of the oil beneath, (from intentional difference of flow in the lamp,) causes the sperm oil lamp flame to fall in brightness, and requires that the wick should be retrimmed. Several causes conspire to this hard condition of the charred cotton, which I need not enter into here.

"I have made many careful experiments on the proportion of light produced by the two kinds of lamps, in every case weighing the oil before and after combustion, so as to know exactly the quantity burnt; and making, during the experiments, above a hundred comparisons of the lights, one with another. The rape oil lamps were always more brilliant than the sperm oil lamps, except, indeed, in one or two rare cases; but, at the same time, more oil was burnt in them. The observations were made numerous, that the errors in the perception of the eyes might compensate each other. In each particular experiment it was evident that the light was nearly in the proportion of the oil burnt; and upon making a comparison of all the results, the following conclusion was obtained: From one hundred and eight observations of the lights, taken at such times as appeared fitted to give the best mean expression of the light of the lamps, compared with the oil burnt in them, the average light of the rape oil lamp came out as one and a half, that of the sperm oil being one. ' This is the mean result of the light involved by the lamps burning for the same period of time. On summing up the amount of oil burnt in the same time, it gave almost exactly the same proportion; for the oil consumed in the sperm lamps being 1, that consumed in the rape oil lamps was 1.505. I have considerable confidence in this result, the quantity of oil consumed being several gallons, and the observations of the light very continuous and numerous."

—

Supplementary return to an order of the honorable House of Commons, dated August 14, 1846.

"Statements by the corporation of the Trinity House of Deptford Strond, by the ballast board in Dublin, and by the commissioners of Northern lights, showing whether they have adopted the recommendation of the select committee on light-houses, on the use of colza or rapeseed oil, instead of sperm oil, and what saving of expense has accrued therefrom; and also whether they have made any, and what reduction, in the light dues charged by them respectively."

Ordered by the House of Commons to be printed, 18th March, 1847.

—

Return by the Commissioners of Northern Light-houses.

1. Whether they have adopted the recommendation of the select committee on light-houses, in the use of colza or rapeseed oil, instead of sperm oil, and what saving of expense has accrued therefrom?

On the 10th January last, the engineer of the board reported on this subject to the commissioners, in the following terms: "At the suggestion of Mr. Hume, M. P., chairman of the light-house committee, I have obtained from Messrs. Wilkins, of Long Acre, a sample of an oil going under the name of oil of colza, but in no respect characterized by the peculiarities which marked the colza oil of the French light-houses. This oil is burned in a lamp with a chimney indented at the flame by means of a deep, narrow groove on its outside, which seems to be the same as that which is used in what is called the solar lamp, and is now oddly enough termed the catoptric lamp. The oil is said to be nearly fifty per cent. cheaper than spermaceti, but is generally held to be more rapidly consumed. I have made several trials of the colza and the sperm oil in burners of different kinds, with somewhat contradictory results, both as regards brilliancy of effect and durability. I am not, therefore, prepared to offer any very precise information as to the result of my inquiry. It appears to me, however, that I may safely enough state that the colza oil is somewhat more rapidly consumed where an equally dense light is evolved, and its flame generally seems to contain a greater proportion of the orange color, which marks imperfect combustion.

"The adoption of this oil would involve an entire remodeling of the burners, so as to admit of thicker wicks, and would probably lead to an enlargement of the cellars, so as to receive a larger supply. I conclude, therefore, that during the present season the commissioners would not be warranted in doing more than to make trial of a cask or two of this oil at a dioptric and a catoptric light. I may add, that other reasons for proceeding somewhat cautiously in regard to the colza oil may be found in the uncertainty as to the source of supply, and the absence of all experimental knowledge of the effect of cold in causing it to become thick."

The suggestions in this report having been approved by the board, a supply of the oil has been sent to several of the stations; but the result has not as yet been reported to the board.

—

A return by the Commissioners of Northern Light-houses, relative to the substitution of rapeseed oil for sperm oil, and the saving that has accrued therefrom.

To the first head of the preceding order, "Whether they have adopted the recommendation of the select committee on light-houses in the use of colza or rapeseed oil, instead of sperm oil, and what saving of expense has accrued therefrom?" the commissioners, in making a return to this order (presented to the House 26th August, 1847, Vetis No. 2) stated that "a supply of the oil has been sent to several of the stations, but the result has not as yet been reported to the board."

On 9th January last the engineer submitted a further interior report to the board, a copy of which is appended hereto (No. 1.)

The engineer having on the 10th instant reported to the board the result of the trials of the oil in question, the same is now submitted as a supplementary return to the order of the honorable House, and is hereto appended (No. 2.)

By order of the board,

ALEX. CUNNINGHAM, *Secretary.*

NORTHERN LIGHTS OFFICE,
Edinburg, March 13, 1847.

—

Excerpt from annual report by Mr. Alan Stevenson, Engineer to the Commissioners of Northern Light-houses, dated 9th January, 1847.

The trials of the colza oil alluded to last year, as suggested by Mr. Hume, M. P., late chairman of the light-house committee in 1845, have just been commenced, and, in so far as the reports from the stations have been received of the preliminary experiments there is reason to believe that this oil will answer in light-houses. The question of the relative expense of the colza and spermaceti oils, however, must remain undecided until further trials shall have been made; and longer experience is also desirable, in order to show in what manner the oil is acted upon by frost. The stations fixed upon for the trials are Inchkeith, Isle of May, Kinnaird's Head, Corswall, Little Ross, and Calf of Man, three being dioptric and three reflecting lights.

—

The report of Alan Stevenson, engineer.

EDINBURGH, *March* 10, 1847.

In my last annual report on the state of the light-houses, I directed the attention of the board to the propriety of making trial at several stations, of the patent colza or rapeseed oil prepared by Messrs. Briggs, of Bishopsgate street. These trials have now been made during the months of January and February, at three catoptric and three dioptric lights, and the results have from time to time been made known to me by the light-keepers, according to instructions issued to them, as occasion seemed to require. The substantial agreement of all the reports as to the qualities of the oil, renders it needless to enter into any details as to the slightly varying circumstances of each case, and I have therefore great satisfaction in briefly stating as follows the very favorable conclusions at which I have arrived:

1. The colza oil possesses the advantage of remaining fluid at temperatures which thicken the spermaceti oil, so that it requires the application of the frost-lamp.

2. It appears from pretty careful photometrical measurements of various kinds, that the light derived from the colza oil is, in point of density, a little superior to that derived from the spermaceti oil, being in the ratio of 1.056 to 1.

3. The colza oil burns both in the Fresnel lamp and the single Argand burner with a thick wick, during seventeen hours, without requiring any coaling of the wick or any adjustment of the damper; and the flame seems to be more steady and free from flickering than that from spermaceti oil.

4. There seems (most probably owing to the greater steadiness of the flame) to be less breakage of glass chimneys with the colza than with the spermaceti oil.

5. The consumption of oil, in so far as that can be ascertained, during so short a period of trial, seems in the Fresnel lamp to be 121 for colza, and 114 for spermaceti; while in the common Argand, the consumption appears to be 910 for colza, and 902 for spermaceti.

6. If we assume the mean of these numbers, 515 for colza and 508 for spermaceti, as representing the relative expenditure of these oils, and if the price of colza be 3s. 9d., while that of spermaceti is 6s. 9d. per imperial gallon, we shall have a saving in the ratio of 1 to 1.775, which, at the present rate of supply for the northern lights, would give a saving of about £3,266 per annum.

Of these conclusions, the three last may be considered as more or less conjectured, being founded on data derived from too short a trial; but the striking agreement of the results obtained at the six lights in which the experiments were made, tend in some measure to supply the place of a longer period of trial; and I have no hesitation, therefore, in recommending the board at once to introduce the use of the colza oil into all the dioptric lights, except that of the Skerryvore, where some special reasons induce me to defer the change for another season. In the catoptric lights, the only reason for not making an equally extensive trial is the necessity for renewing all the burners, which require to be so constructed as to receive thick wicks of brown cotton; and it may perhaps be considered prudent to proceed with some caution in changing the apparatus so as to suit for burning a patent oil, the circumstances attending the regular and extensive supply of which are not yet fully known. I may remark, that I have burnt the colza oil in the solar lamp alluded to in my last report; but I disapprove of it as tending to elongate the flame vertically, and thus to decrease its horizontal volume. The elongated form of flame increases the divergence vertically when the light is lost, and so far circumscribes its horizontal range where it is most required. I have, therefore, substituted the thick wick burner for the solar lamp, whereby an equally complete combustion is obtained, and the proper form of the flame is at the same time preserved.

To the BELL ROCK COMMITTEE.

Letter from W. C. Wilkins, Esq.

LONDON, *July* 18, 1850.

SIR: In answer to your letter of 29th of May, requesting information concerning improvements in light-houses and light-vessels, I beg herewith to enclose a book which I have recently published containing drawings and particulars of the various catadioptric light apparatus on Fresnel's system, used in this country, including two apparatus for which I have obtained letters patent in connection with M. Letourneau, of Paris; one is a revolving light, short eclipses produced by a new arrangement of the lenses, of the central part and additional vertical prisms, gathering the rays from the upper and lower catadioptric portions, giving a great increase of light, and is what I am now exhibiting in the great crystal palace, Hyde Park; the other is a reciprocating apparatus, with additional vertical prisms before mentioned, and reduce the cost of flashing light when it is not required to be seen beyond the same circle, or half the horizon.

I have also effected great improvements in revolving floating lights by placing the machinery below the deck, thereby avoiding the liability to injury, and producing a better action. I have enclosed circulars of these novelties, to assist in conveying the idea of their advantage to you.

In reply to the latter sentence of your letter, seeking intelligence respecting the oils, I beg to say that the colza or rapeseed oil is in constant use here; it has quite superseded sperm, and with suitably constructed lamps, wicks, &c., gives a better light and is more economical. (I shall be happy to supply you with a sample.) After it has been pressed from the seed it is purified by a patent process, which is of course secret—known only to the parties who manufacture it; the price averages about 3s. 10d. per gallon, but fluctuates according to the demand. Numerous experiments have been tried under the direction of Professor Faraday, and he considered the rapeseed oil as superior to all others for the purpose of illuminating light-houses, &c.

Trusting that this will meet your approval and contains what you wish to know, and that I may be honored with your future commands, which shall always have my best attention,

I am, sir, your most obedient servant,

W. C. WILKINS.

Lieut. T. A. JENKINS, U. S. N.,

Secretary to Light-house Board, Washington.

REFUGE BUOY-BEACON.

The ordinary buoys in use on the English coast and elsewhere, for pointing out the position of dangerous shoals and sand banks, are of a conical form, made chiefly of wood, and hooped like a cask, being moored with the apex or sharp end downwards; and, owing to this shape, in a strong tide-way and heavy sea they are at times nearly pulled under water, tugging with an immense strain upon their moorings, and frequently breaking adrift at the very time when most required; moreover, from their construction, they twist and twirl so as to render them impossible of approach or refuge for saving life in case of shipwreck. Captain George Peacock, the superintendent and dockmaster of the Southampton docks, has invented a new kind of mark-buoy, or floating-beacon, which, from its peculiar form and construction, and the manner in which it is moored, rises over the *crest* of the waves in the heaviest gales and strongest tide, and is not acted upon like the ordinary buoy; it is, in fact, capable of holding from ten to twelve persons with ease in case of contiguous shipwreck, and of affording them a safe temporary asylum—it being in fact at once a buoy, beacon, and life-boat, and the cost of it scarcely exceeds that of the ordinary buoys now in use.

The plan and model having been approved of by the pier and harbor commissioners of Southampton, one of these buoy-beacons was laid down off the Spit, at Calshot Castle, on the 12th of August last, and has withstood the whole of the severe gales of the last six months without showing the least symptom of injury or leakage.

The hull, which is of sheet iron, is of a semi-oval shape, like the horizontal half of an egg, being ten feet long, seven feet broad, and three feet deep: the deck, a perfect oval, is convex, with a man-hole and cover in the centre. A kelson of pine timber, eighteen inches deep and fourteen inches broad, is fitted to the inside, running fore and aft, and fayed on to the shape of the bottom upon the rivets; and the mooring-bolt, with a broad bearing-shoulder, is passed diagonally up through this kelson and firmly secured by a large nut over a plate of iron or washer on the top of the same. This mooring-bolt is also fixed at one-third the length of the hull from the large end or breast, and along the other two-thirds of the bottom an iron keel, fifteen inches deep and 250 pounds weight, is rivetted on with angle iron, and bolted through the kelson, which keeps the buoy steady to the tide, and also gives it stability. To the side or rim, which is eighteen inches deep and inclined inwards, uniting the deck with the bottom, eight triangular-shaped wooden uprights are fixed at equal distances in outside sockets bolted through with nuts and screws: these stancheons are nine feet in length, terminating and dovetailing into an oval platform, five feet by four feet, and are braced horizontally by two rows of corresponding pieces at equal distances from each other, the first row being four feet from the deck; and the divisions above this are nearly filled up by the vertical battens to the top of the platform, all firmly united with hoop iron. There are also two diagonal braces of rod-iron, which cross each other in the centre of the structure; whilst seats are fixed at each end on a level with the first horizontal brace pieces, affording accommodation for six persons, and leaving standing room for six more in the centre of the deck. Under the platform a large bell is fixed, with four swinging clappers hanging round it from the platform and striking the outside; whilst the centre clapper has its stem below the hammer elongated with rod-iron to five feet, terminating in a wind-cross of thin sheet iron, so as to ring the bell with the least breeze when the water is too smooth to affect the clappers.

Above the platform, arching from side to side, is a semi-circle of square iron rod, which assists in uniting the sides or top ends of the uprights or stancheons to the platform, and to which rim is rivetted a plate of thin iron, with the name of the buoy painted on it; and above the centre of this arch, forming the apex of the buoy-beacon, is a spindle carrying a pyramidal speculum, ten inches in angle, which, revolving freely as the buoy moves, reflects the rays of the sun and moon, and occasionally Calshot light. The reflected flashes of the sun's rays are visible at a distance of seven to eight miles from a vessel's deck, and the buoy-beacon itself is seen in clear weather four to five miles off, or three times the distance of ordinary buoys of the largest size; in thick weather it looms like a small vessel at anchor. The top of the speculum is twelve feet above the water-line.

The commanders of steamers and other vessels frequenting the port of Southampton, and all the pilots, give a unanimously favorable report of this buoy-beacon, and strongly recommend its general adoption upon all outstanding dangers; they say that it rises over the tops of the seas without plunging or diving, or being in the least affected by the action of the tide beyond sheering from side to side within the scope of its mooring chain during a gale across the tide, and thus rendering itself more conspicuous by presenting two-thirds of its broadside each way alternately; and in heavy gales of wind from the most exposed quarters, namely, SE. and WSW., at the strongest period of a spring tide, it is seen to ride upon the crest of waves that completely overwhelm the neighboring buoys of the Bramble, Leep, &c. It has not been baled or pumped out since it first went into the water, and upon taking off the manhole-cover, after it had been laid down six months, it was found to be as tight as a cup. As a proof of its stability, two persons at one time have sat on the top of the platform when afloat, without its showing any signs of being crank. The cost of this buoy-beacon, including bells, fittings, &c., was £55.

Letter from Jacob Herbert, esq., Secretary to the Trinity House Corporation.

TRINITY HOUSE, LONDON,
September 3, 1851.

SIR: I beg to acknowledge the receipt of your letter, dated 27th May last, stating that you are directed by the Light-house Board in session in conformity to a recent act of Congress (copy of which is therewith transmitted) to address this corporation, for the purpose of obtaining information which it is considered they may be enabled to furnish, either of a general or specific character, with a view to the improvement of the present condition of the lights and other aids to navigation belonging to the Government of the United States; and thereupon calling the attention of the Elder Brethren to certain particular inquiries deemed of major importance, embracing a period of time subsequent to the date of the House of Commons's report of 1845.

And having laid the same, with its accompanying printed documents, before the board, I am instructed to communicate to you, as secretary of the Light-house Board of America, the following replies which the Elder Brethren have directed to be given to the queries referred to, in the order in which they occur in your said letter, and which I trust will convey the desired information.

I have the honor to be, sir, your most obedient servant,

J. HERBERT.

To Lieut. THORNTON A. JENKINS,
U. S. N., Washington City, U. S.

Q. As to the changes which may have been made in the mode of constructing light-house towers; the different descriptions of materials used, which is considered the best under all circumstances, referring specially to durability, economy, and efficiency as permanent structures, having regard to annual expense for repairs?

A. The mode of construction remains unchanged, except that, as respects material, iron-plate houses have been erected in some of the British colonies. As a general rule, stone is preferable, but much depends on locality and other circumstances.

8

Q. What has been the experience of the board in the use of the screw-pile and Potts's cylinders for special locations where hydraulic structures would have been attended with great expense or difficulties which could not have been easily overcome? What objection, if any, to cast-iron hollow towers in general? What has been the effect of vibration and torsion of the pile structure and of vibration upon the hollow cast-iron towers, so far as the Trinity board has been informed?

A. Neither the screw-pile nor Dr. Potts's cylinder have been extensively used by the Trinity House. The vibration of the pile structures is sensibly felt. No torsion has been observed. The board has had no experience of cast-iron hollow towers.

Q. The results of trials of mortars and cements for stone and brick towers; the effect of the sea atmosphere on hydraulic and other light-house structures?

A. These are points for the opinions of engineers.

Q. Improvements, if any, in illuminating apparatus since 1845?

A. The dioptric lights have been improved by the substitution of zones of prisms for concave mirrors.

Q. What number of new lens or Fresnel lights? What number of old reflector lights renewed by the introduction of the lens apparatus, and what the whole number of lights introduced since 1845? Have lenses of the smaller classes been introduced for harbor and river purposes?

A. Three new lens lights have been established, viz: one at Trevose Head, and two in Sea Reach. The Milford high, Eddystone and Spurn low lights have been renewed in the manner mentioned; a catadioptric apparatus of the first, second, and fourth orders, respectively, having been substituted for the Argand lamps and parabolic reflectors theretofore in use thereat. Lenses of the fourth order have been introduced and still are in use for harbor and river lights.

Q. What the comparative useful effect and economy of that class or order of lights, with the reflector lights in similar positions? What the relative economy and useful effect of the two systems as demonstrated by experiment at the South Foreland lights and other places in a series of years, all things else being equal?

A. The comparative economy depends upon the number of lamps required under the old or reflective system.

Q. What lamp is in most general use in the lens lights, the Carcel, pneumatic, or hydraulic lamp? The advantages arising from the use of the one or the other?

A. The hydraulic or fountain lamp is universally in use in the light-houses of the corporation, with one exception (the South Foreland high) where the light is shown from a Carcel lamp; the disadvantage arising from the use of which is the occasional derangement of the machinery.

Q. What the most approved lamp for reflector lights?

A. An Argand lamp: the burner seven-eighths of an inch in diameter.

Q. What improvements in ventilation? Have Prof. Faraday's tubes been generally introduced?

A. Prof. Faraday's tubes have been fitted at all the light-houses.

Q. How often do the best quality parabolic reflectors in use in the Trinity House lights require renewing?

A. Not for many years with care—say from twenty-five to thirty.

Q. How long do the best slide lamps and burners last with ordinary repair?

A. The best Argand burners, when new, last from two to three years; the lamps a much longer period.

Q. What is the difference in first cost and what the difference in annual expense of the two systems, comparing lights of the same class or order of the two systems in which other circumstances are equal? What the difference in useful effect, &c., &c.?

A. The difference in first cost is in favor of the reflector lights, as follows, viz:

For the dioptric apparatus complete, about £1,030
Lantern, pedestal, &c., &c 1,940
                                                  ——— 2,970
Reflecting apparatus complete, about 388
Lantern, pedestal, &c., &c 1,940
                                                  ——— 2,328

Difference £642

Q. Have any difficulties attended the management of the Carcel, or other single burner lamps, since the introduction of the lens lights into Great Britain?

A. No difficulties have attended their management; the only disadvantage is pointed out under the above reply to the third query.

Q. What difference is made in regard to numbers, salaries, and qualifications of keepers in the two systems?

A. No difference is made in these respects.

Q. Have any improvements been made in the illuminating apparatus of light-ships? Are Argand burners and parabolic reflectors mounted on gimballs used on board the Trinity House light-ships?

A. Argand burners and parabolic reflectors mounted on gimballs are used on board the Trinity House light-vessels.

Q. What kind of fog-signals are employed?

A. A gong is sounded.

Q. Have any new modes of distinguishing lights been introduced? and what are now employed?

A. No new modes have been introduced. Distinctive character is obtained by varieties of fixed, revolving, intermittent, and colored lights.

Q. What number of persons constitute the crews and officers of light-vessels in exposed positions, and what number in rivers, bays, &c., &c.; what the tonnage of the light-vessels on the Goodwin sands and in other exposed situations?

A. Eleven persons in all, consisting of a master, mate, and nine seamen, excepting in one instance, in which the number of men is fifteen, (seventeen hands in all.) One hundred and sixty tons, more or less, by builders' measurement.

Q. What descriptions of oils are used in the lights of England?

A. Colza or rapeseed, refined.

Q. How are they tested before being received?

A. Tested by experimental burnings.

Q. What number of light-vessels kept for relief?

A. One in each district; the number of floating light stations in the several districts varying from eleven to three.

Q. What has been the result of trials with colza or rapeseed oil, chemical and gas lights, etherial oils, &c., &c.? Do you use different qualities of oils for winter and summer?

A. The result, at present, is the invariable use of refined oil from rapeseed, as before stated. No other oil is used.

Q. Of what materials are lanterns, in general, constructed for the Trinity House lights?

A. Lanterns are constructed of copper or gun metal.

Q. What advantages arise from the use of bronze?

A. Bronze, like gun metal, does not corrode.

Q. What description and sizes of glass for seacoast lights?

A. Plate glass : the dimensions of the panes vary. In the first order lights from 48 by 41 inches to 35 by 29; second order lights from 41 by 39 inches to 22 by 19. The thickness of the glass being ordinarily three-eights of an inch; in exposed situations five-eighths of an inch.

Q. How is the glass put in, by being glazed with putty, or by other means?

A. The panes are fixed by means of a composition of white and red lead termed plate-glass cement.

Q. What precautions are observed to repair such casualties as the breaking of glass by wild fowl, mechanical lamps, or movable machinery getting out of order, &c.?

A. No precaution is taken against the breaking of the glass by wild fowl, but spare panes are kept in store; instances of breakage are very rare. Movable machinery, &c., seldom gets out of order, and no particular precaution is therefore considered necessary.

Q. What precautions are taken to prevent unnecessary waste of light by absorption in the lanterns?

A. The inner part of the lanterns is painted white.

Q. Are seacoast or other first class lights permitted to be left after lighting up at sun-setting and before being extinguished at sun-rising, without a keeper in the lantern?

A. No. A keeper is always in the watch-room. (Vide printed instructions to light-keepers.)

Q. What is the general system of inspection? How often, and by what class of persons?

A. By superintendents of districts, and frequently by members of the board.

Q. How are supplies delivered—by tenders, or by contract vessels? How often per annum, and under whose direction, &c.?

A. By the corporation's vessels once a year; or oftener, if necessary, under the board's order.

Q. What the comparative expense of the Maplin sand screw-pile light, with a first class light-ship, per annum, and which is the most efficient?

A. The Maplin screw-pile light in the year 1849 cost, say £455. The light-vessel at the north end of the Goodwin sand, say £1,048.

Q. Any information on the subject of buoys, beacons, and sea-marks, with reference to inspection, distinction, &c., is desired.

A. They are constantly inspected, and are distinguished by difference of color and character.

Q. Modes of giving publicity to proposed changes in regard to lights, &c. How long before the change is made, and what specific means employed to give extended circulation at home and abroad ?

A. By advertisement in the principal daily newspapers of the metropolis, &c., in a few of the periodicals generally perused by mariners. Separate notices are also fixed up and distributed at the different custom-houses, and copies sent to the consuls-general of the various foreign governments likely to be interested in the notifications. In some cases preliminary notices are issued.

Q. Any printed forms, circulars, reports, descriptive lists, general and special instructions, &c., upon the subject of lights, buoys, and beacons, under the Trinity House corporation, will be most acceptable.

A. General instructions to light-keepers : general instructions to masters and mates of light-vessels ; directions for the management of refracting and reflecting lights to light-keepers ; directions for the management of reflecting lights on board light-vessels; various forms of account of oil, &c.; inventory of stores for a light-vessel.

Extracts from the report of the Hon. R. J. Walker, Secretary of the Treasury, August 5, 1846. (Senate document, No. 488, 1st session, 29th Congress.)

[Report of Lieutenants T. A. Jenkins and R. Bache, United States Navy.]

" In countries where the system of lighting is good, a general plan for the classification of lights, selection of sites, construction of the light-houses and apparatus, and inspection, has been adopted, or improvements have been slowly introduced under the direction of existing authorities. The most perfect system in Europe is the result of the former plan; and it must be obvious, that, for a growing country like ours, no other can secure, on the greater part of the coast, a due attention to the wants of commerce and navigation.

"The unequal distribution of lights upon different important parts of our coast proves this, and shows how desirable it is that some general plan should be fallen upon, by which the claims of different parts of the country might be duly considered, a light provided wherever it is desirable to navigation, and expenditure saved when no such necessity exists; that, when a light-house was decided to be necessary, the site of the buildings should be properly selected, the buildings properly constructed so as to combine stability and economy; all the accessories, as store-houses for oil and for lighting apparatus, for fuel, and the like, properly arranged; that the lighting apparatus should be of the most approved kind and construction, and so placed as to illuminate the required part of the horizon without waste; that the lights should be properly classified, so that, from the large seacoast light to the small harbor light, the buildings, lighting apparatus, and accessories might be on the proper scale; the distinctive character for the lights adopted which experience has shown most effective, as fixed, revolving, fixed lights with flashes, successions of bright light and eclipses at regular intervals, and lights of particular colors; that when ready for illumination, the lighting apparatus may be properly kept and attended to, under effective inspection.

" It is not possible to leave such things as these to local information, derived at second hand, or to chance, without paying dearly in the end for the want of system.

"In all these points it appears that other countries have advanced rapidly; and it becomes us to see what improvements introduced elsewhere are adapted to our use, and to ascertain how our general system may be modified.

"A comparison of the lighting apparatus in use with us, and of the improved apparatus of France, shows that in this essential our progress has not been as great as might have been desired.

"The reports of the inspecting officers detailed from the navy to examine the lights on our coast showed their absolute defects; the present report shows their deficiencies relative to those of other countries. The trial made of one of the French lights, at the entrance to New York harbor, at Sandy Hook, has been very successful, but the use of this apparatus has not been extended. A light-house need be seen from particular parts of the horizon only; and as the lamps used throw off their lights almost equally on all sides, the direction of portion of the rays must be changed. This may be done by reflection from opake substances, as from metallic mirrors; or by bending or refracting by transparent bodies, as by glass prisms or lenses. Much more light is lost by the use of the most highly polished surfaces as reflectors, than by passing through transparent bodies; and hence lenses are more useful and economical for light-house purposes than mirrors. The refracting or lens apparatus invented in France by Augustin Fresnel, has been generally applied under the direction of his brother, Leonor Fresnel, the present accomplished secretary of the Board of Light-houses.

"In this apparatus, lenses of glass are used to throw the light of powerful lamps in the required directions upon the horizon, and the part which would otherwise escape upwards is reflected to the horizon by glass prisms, so placed that the rays fall upon the back or second surfaces of the prisms, at angles at which they are entirely reflected. The lenses are built up of separate pieces of glass, thereby saving the weight and cost of large masses of material, besides diminishing the thickness, and thus preventing the loss of light by absorption, and permitting the figure of the surface to be adjusted so as to bring the rays accurately to a focus. If the lens apparatus be made to revolve, the navigator will see brilliant flashes of light, gradually growing dim, and succeeded by short intervals of comparative darkness; and if a second set of lenses, suitably arranged, revolve about the first, which remains fixed, a general illumination will be varied by brilliant flashes; these may be white or colored; and thus the appearance of lights may be so varied as to make them distinctive. A single large lamp, supplied with oil by mechanism, (as in French light-houses,) or by hydrostatic pressure, (as in some of the English ones,) is used with the lens system, and produces great economy in the consumption of oil necessary to supply a given quantity of light.

"The dimensions of the apparatus are easily varied to suit the different quantities of light required, from the small harbor light to the large seacoast light. In the French system, the lights are divided into four orders, the first being the most powerful; and each order may be sub-divided into a larger or smaller size. The best reflecting lights do not more than reach, in power, the second order of the lens system.

"The advantages of the Fresnel or lens system are: 1st, in the greater brilliancy of light; 2d, in the greater quantity of the more brilliant light thrown upon the horizon; and 3d, in the less consumption of oil in obtaining these advantages. Simple experiments and calculations based upon them give an unerring means of ascertaining the relative brilliancy, quantity, and economy of different lights. If two lights are equally bright, they will illuminate equally two equal surfaces placed at the same distance from them. If one is brighter than the other, the brighter will illuminate equally the same surface when farther from the light; a fourfold brightness corresponding to a double distance, a ninefold brightness to a triple distance, and so on. Thus it is easy to compare the relative brightness of lamps, and adopting the light of some particular lamp as the standard, to describe other lights as equivalent in brightness to once, twice, or more times the standard.

"The useful effect of a light depends not only on its brightness, but on the extent of horizon which it can illuminate. The average brilliancy in different directions, multiplied by the extent of horizon over which it shines, gives its useful effect. Its economy is measured by this useful effect directly and inversely by the consumption of oil required to obtain it.

"The careful experiments of Mr. Fresnel leave no doubt of the great advantages in respect to brilliancy, useful effect, and economy of the lens system, as may be seen by a few examples.

"A harbor light on the lens system, of the smaller size of the fourth order, with a single mechanical lamp burning $1\frac{9}{16}$th ounce avoirdupois of oil per hour, would give twice as brilliant a light as the ordinary reflecting system having a lamp burning $1\frac{1}{4}$ ounce avoirdupois of oil per hour. The quantity of light on the horizon would be double, the cost of a given quantity of light one-half, and therefore the economy two-fold. As the apparatus increases in size from this to the higher orders, the advantage of the lens system increases. In the third order, second size, a mechanical lamp with a double wick, burning $6\frac{3}{4}$ ounces avoirdupois of oil per hour, gives as much light as fourteen lamps with reflectors, *each* burning $1\frac{3}{4}$ ounce avoirdupois of oil per hour. The useful effect is one and a half times, and the economy between three and four fold. In the second order of lights the new system for equal useful effects is three times as economical as the old, reaching in the larger sizes to four-fold. A power equivalent to that of the first order of lens lights has not been reached by the reflecting system.

"The cost of the erection of buildings for the new system of lighting, and the first cost of the apparatus itself is somewhat more considerable, and the number of keepers required is greater than with the old; the repairs, on the contrary, are much less. An accurate comparison of these particulars shows that in France the economy is in favor of the new system. Thus, taking into consideration the interest on the cost of tower, lantern, reflectors or lenses, and of keeping up the light, the relative expense of the two plans for a small harbor light is as 236 to the new to 226 of the old plan; while the quantity of light on the horizon is as 2 to 1 in favor of the new; and hence the economical effect is nearly double upon the new system. For a large revolving light (second order) the annual outlay for the old and new systems would be as 126 to 208, while the useful effect would be only as 1 to 2, and the economical effect of the new system would be more than one and a half that of the old. The cost for the lens apparatus might for the present be greater in our own country; but the economy in lighting by the lens system is too great for this circumstance to turn the balance against it.

"The mechanical lamp used with these lights, or some other which American ingenuity may supply, or the hydrostatic or pneumatic lamp in use in the English light-houses, will replace with advantage the present imperfect lamp. In France, the mechanical lamp is found to require but small repairs, readily made in establishments where the lamps are constructed; and both construction and repairs would surely be practicable here. The cost of repairs of the lens apparatus is in a series of years merely nominal, and experience has shown that it is more secure and more easily seen than the old. No important seacoast lights should be left without being watched by a keeper, and, in the economy in lighting, will much more than pay the cost of two keepers in the larger light-houses.

"Whether the rapeseed oil generally used in the French light-houses may be employed in our own with advantage, is a question which cannot now be settled; it may, however, be desirable to call the attention of farmers to the cultivation of the plant from which it is obtained.

"The use of the screw-pile, for the foundation of light-houses, has, by rendering the establishment of permanent structures upon banks and shoals comparatively easy, safe, and economical, superseded in many cases the use of light-boats, which, especially in exposed positions, are of comparatively little value.

' The buoys used in the entrances to our harbors are now placed by local authorities, and under loose regulations. A general system should be adopted of coloring and numbering, and should be so rigidly adhered to that the seaman would know his position as soon as he discovered a buoy. This is practicable, as will be seen from the interesting account, in the report of Lieutenants Jenkins and Bache, of the intricate approaches to the port of Liverpool, which are rendered quite safe by the system of buoys, lights, marks, and tide signals. The natural marks which disappear yearly from our coast should be replaced by permanent ones; screw-piles for mooring buoys should in certain cases be supplied. The arrangements for placing buoys and verifying their positions require to be rendered systematic, and to be subjected to some general control. The navigator should have due notice of all changes from a source connected with the whole light-house system.

"The best modes of lighting would be ineffective unless the keepers were careful and intelligent persons, and instructed in the necessary particulars of the business. Some training is desirable. Frequent reports, and perhaps the suggestion in regard to keeping meteorological and tide registers, as a means of securing attention and intelligence, may be adopted."

*   *   *   *   *   *   *

Extracts from Senate Document No. 488, 1st session, 29th Congress.

[Report of Lieuts. T. A. Jenkins and R. Bache, U. S. N., to the Secretary of the Treasury, August 5, 1846]

LIVERPOOL BAY AND HARBOR.

The buoys are constructed of both wood and iron. Those of iron are being pretty generally introduced by the surveyor. Six of the largest employed are of that material, and he expresses a decided preference for them, both in an economical point of view and as a matter of expediency. They are constructed with water-tight compartments, so that there is no danger of their filling should they be injured by a vessel coming in contact with them. Galvanized iron, both for buoys and for the hoops of wood ones, is highly approved of. The expense is the great obstacle. For buoys to be painted white, there can be little doubt of the great advantage of using the galvanized iron for hoops. Can and nun buoys are employed. (See accompanying drawings.) The iron buoys require a weight to each to make them float properly in the water, but, once properly ballasted, there is no further difficulty with them in that respect. The buoys are of good sizes, fine proportions, and are distinguished by their color, shape, number and name.* Here, as in all the ports of the kingdom, red buoys are placed on the starboard hand of the channels leading from seaward, and the black ones on the port hand. The numbers commence from seaward. At the end of spits, or at turning points, perches are placed on the buoys (of iron.) So well and perfectly are the buoys distinguished that a man, to go wrong, must be unable to read, or to distinguish colors.

The moorings of the buoys are heavy iron sinkers, so moulded as to increase their tenacity of hold, with the heavy chains sent from the light-vessels as unfit for use, although perfectly good and applicable to this purpose.

There are in the store duplicates of all the buoys; they are painted regularly once in six months, and their moorings raised every year.

The buoys and light-vessels are placed in their proper positions by the surveyor himself, the positions being fixed and decided upon by angles, so that the exact spots are known upon which to place them, and from which they are never allowed to be removed, unless alteration of shoals or channels make it necessary, (when it becomes necessary to consult the Trinity corporation, London,) or by accidents occurring to them; in which latter case a few hours only are allowed to elapse before they are replaced.

Registers of the buoys are kept, (see form,) showing when they were made, when repaired, and when placed, &c., with a column showing their positions by angles, remarks, &c. The present courteous and intelligent marine surveyor of the port of Liverpool, (Lieutenant Lord, R. N.,) is greatly in favor of iron as a material, as well

---

* Thus: "*Red* buoys on the *starboard* hand, and *black* on the *larboard*, when running in."

"*Black* and *white* striped buoys on intervening banks or flats."

Superior can buoys, with perches at the "*elbows* or *turning points* of principal channels."

"Each buoy bears the *initial* of the channel it occupies, thus: F. *Formby channel.* C. *Crosby channel.* HF. *Half-tide Swatchway;* N. *New channel.* H. *Horse channel.* R. *Rock channel.* HE. *Helbre wash.* B. *Beggars' Patch.* L. *Hoylake.* V. *Victoria channel.*"

'The buoys are likewise *numbered* in rotation, No. 1 denoting the outer or *seaward buoy* of the channel its letter indicates."

for light-vessels as for buoys. The Northwest light-ship is 203 tons, and cost no more than one of the same size of wood would have cost, while the advantages in many respects are greatly on the side of the iron. The larger the vessel the greater the economy of iron in comparison to wood. They draw but little water, (the Northwest only nine feet,) and are fitted with bilge keels to keep them from rolling too heavily. The greatest possible attention is paid to all moorings. Experiments have been made upon chains of different sizes, &c. Those in use at present are fitted with wrought-iron studs, and have been found to answer infinitely better than those with cast iron ones; the former seldom or never falling out.

This corporation, not satisfied apparently with this perfect system of lighting, &c., has further provided for the safety of seamen by adding a number of life-boats to their charge. We are aware that this does not come strictly within the limits of our instructions; but finding it intimately connected with the light-house establishment, and having heard of the many and great benefits which have resulted from it as being a part of the most complete whole, we deem it but proper to refer to it in this connexion.

The boats (nine in number) are placed at different stations around the bay, well provided and well protected from the weather, ready at a moment's notice for service. The crews are composed of experienced watermen residing near the stations of the boats, always willing and anxious to render assistance to those who may be so unfortunate as to strike upon the shoals, or meet with any other accident within the vicinity of the bay and harbor.

To insure a more certain and prompt assistance to those in distress, the corporation has an arrangement with the Steam-tug Company, by which, for the consideration of 400 guineas per annum, and 25 guineas additional for every time a steamer is required, a vessel is ready at all times, day and night, with steam up, for towing out a life-boat, and for rendering any service that may be required.

A simple inspection of the chart of the bay of Liverpool will suffice to explain the manner of masking the lights, by which the positions of the buoys are pointed out by night to the pilot, so that he can stand on his course with confidence. The system is simple and perfect, meriting the warmest eulogies of all persons feeling any interest in the commerce of the port.

LIGHT-HOUSES OF SCOTLAND.

The " Commissioners of Northern Lights," who are charged with the control and management of light-houses, buoys, beacons, &c., of a general character, on the coasts of Scotland, were first instituted by act of Parliament in 1786.

The board is composed of twenty-five commissioners or members, whose services are gratis, and whose professions are of a civil character, but chiefly that of the law. There is not a single nautical person attached to the board.

There is a secretary to the board and, an engineer who is the executive officer ; an auditor, and an accountant, who prepares all accounts for the committees, &c.

The local or harbor lights, fifty-one in number, are under the control and management of local authorities and trustees, supported by dues levied upon the shipping visiting the respective ports where the lights are established.

The general board meets four times a year for the transaction of business. It receives and acts upon the proceedings of the committee meetings which have been held in the interim.

There is a standing committee, composed of sixteen members, called the "Bell Rock committee," which meets once a fortnight during the year, except in the months of April, August and September. This committee has charge of the lights, and appoints the finance committee, before which the auditor is required to lay all accounts twice a year, when they are examined and passed.

There is also a committee regulating "superannuated allowances," and one for "buoys, beacons," &c.

Committees are appointed for the superintendence of light-houses in the course of erection ; and the board is in the habit of remitting any matters of importance which may arise to the consideration of a special committee, which examines and reports upon them.

The engineer of the board attends upon all the committees. He executes all orders relating to lights, buoys, beacons, &c., and has the entire management of the executive department. He receives a salary of £700, and an allowance of £200 additional for clerk hire.

There are no agents for the inspection of lights. The engineer goes around to all the stations at least once a year, and reports the result of his observations to the board. There are two persons in the engineer's department: the "superintendent of light-keepers' duties," who is charged with the delivery of stores at each station, accompanying the tender which delivers them. (which takes place generally twice a year,) reporting, on leaving the station, the condition in which he found the establishment; and, on his return, is required to hand over to the engineer the diary of his journey, in which is entered the observations made by him upon each station. The other person is the "foreman of light-house repairs," who, assisted by one or two men, goes, as occasions may require, to repair the lamps and light-house apparatus in general.

The engineer makes out a requisition once a year. according to a printed form, (hereto annexed,) for all stores that may be required for each light-house. These requisitions are submitted to a sub-committee, which rejects such items as may not be satisfactorily explained by the engineer, and approves all those deemed necessary for the service.

Printed forms are then issued to different parties, generally in Edinburgh, one of which each party retains, and returns the other, with the prices attached to each item required, accompanied by patterns or samples. The lowest tenders are always taken in the cases of ordinary supplies. There are no public newspaper advertisements for supplies; but applications are made to parties who are considered able to undertake the contract.

The supplies are delivered at the storehouse at Leith, compared by the engineer, the superintendent of light-keepers' duties, and the foreman of repairs, and, if found to correspond with the requirements of the tender, are received in store, from whence they are forwarded, in the tenders belonging to the establishment, to the different stations, as before mentioned. There are at present two light-house tenders, and it is proposed to build another, to be propelled by steam.

All applications for new lights, &c., are submitted to the standing committee. Examinations are made for proper sights, &c.; and, if deemed necessary, after having obtained the approbation of the Trinity Corporation of England, proceed with the work. The engineer presents the plans, specifications, &c., which must be approved by the board before the commencement of the work. The light-houses are generally built by contract, under the immediate direction and superintendence of the engineer of the board, and a superintendent who is not permitted to leave the spot, after the commencement of the work, until it is completed and received by the board. All extraordinary repairs are made in the same manner.

The buildings are constructed chiefly of the dark stone or granite peculiar to the coasts, in a plain, substantial manner. The oil cellars are fitted and finished with a proper care, as in England and France : the keepers' houses are separate from the towers, and differ very little in their general arrangements from those of England. One very simple and useful custom of calling the keepers from their dwellings to the towers, a distance ordinarily of forty or fifty feet, was observed, which removes all necessity for the keeper to leave the light-room until regularly relieved. This consists of a small metal air-tube, leading from the lantern to the room of the keeper, at the end of which are bells, which are struck by small hammers of wood, raised by blowing into the tube. The keeper, upon being called, answers in the same manner ; the one making the call reversing the hammer of his bell after making the signal, so that the answer may be made.

The light-keepers' houses are furnished with spare rooms for the engineer, and for the workmen who visit the stations for making any repairs that may be required upon the illuminating apparatus. The furniture of these rooms is supplied by the commissioners.

The domes of the lanterns are all double-roofed of copper. No lightning rods are used.

The reflectors are parabolic, and generally twenty-one inches in diameter for the fixed lights, and twemty-four inches in diameter for the revolving lights.

The lights are distinguished very much as they are under the Trinity board, with some little modification. In a few cases there are two lights, placed one above the other, in the same tower—a plan not to be approved of under ordinary circumstances. The number of burners varies from one to forty-eight for each station, or from one to twenty-seven for each light-house; the average for each light being about twenty.

The lamps employed in the dioptric lights are the same as those employed in France. Those employed in the reflector lights are Argand fountain lamps, with burners containing wicks of about one inch in diameter. Great care is bestowed upon the manufacture of these lamps, which have their burners tipped with silver. to prevent their too rapid destruction by the great heat of the flame produced by them.

The Argand lamps are solidly made of brass, and of different forms; although those most modern, most approved of, and most in use, are fitted with a slide apparatus accurately formed, by which the burner may be removed from the interior of the reflector at the time of cleaning or wiping it, as also for trimming the lamp, and returned to exactly the same place, and locked by means of a key. The arrangement is an admirable one, as it insures the burner always being in the focus, and does not require that the reflector be lifted out of its place every time it is cleaned. The reflectors are securely screwed to the frame, and the focal points marked upon them for the flames of the lamps. The lamps are made in Edinburgh, and the reflectors in Birmingham, as a general rule. Lamps are also made in Birmingham. Great attention has been paid to the ventilation of the Scotch light-houses. Professor Faraday's tubes are highly approved of, although they increase the consumption of oil.

The engineer to the Board of "Northern Lights" has the entire management of the lights in the executive department. He selects his assistants to act under him in the construction of buildings, and visits the lights at least once a year. They are also visited by the superintendent of the light-keeper's duties, and by the foreman of the light-house repairs.

The light-keepers are appointed by the board. Should they fail in their duty, the engineer reports them to the board. There are two keepers at each light-house—one a principal, with a salary of £50 per annum, and an assistant, with a salary of £40 per annum. At the Bell Rock and Skerryvore there are one principal, one principal assistant, and two assistants; with salaries, for the keepers, of about £70, principal assistant £65, and the assistants each £60. In addition to their salaries, they are allowed each a suit of uniform clothing or watch cloak once in three years. They are also allowed a piece of ground large enough to produce grass for a cow, and a small garden. New keepers are instructed in their duties for three months.

In addition to the returns required of the keepers of their expenditures of oil and other supplies, they are required to keep a barometrical and thermometrical journal, with remarks upon the winds and weather.

They are suppled with timekeepers, and their dwellings are kept well painted and repaired; they were in good order.

There are no light-vessels on the coast of Scotland.

The beacons and buoys are under the immediate direction of the "committee;" under the direction of which they are placed, inspected, repaired, &c. The buoys are made of wood, and examined and painted twice a year.

"Since the rule for exhibiting the lights between 'sunset and sunrise' has been adopted in Scotland, an increased expenditure of oil has been occasioned. This increase, if we estimate four hundred and ninety-two burners at five gallons more throughout the year, which seems a fair allowance, is two thousand six hundred and forty gallons, or about £786 13s. 9d. per annum. The former rule, of making the 'going away and return of daylight' as the times of lighting and extinguishing, was departed from with a view to uniformity of practice."

FRENCH LIGHTS.

The light-house department of France is attached to the official duties of the minister, Secretary of State for the Interior, and is under the immediate control and direction of the Minister of Public Works, charged with the administration of the bridges and roads.

A central public board has the management of all light-houses, buoys, beacons, and sea-marks on the coasts, which is composed of eleven distinguished scientific and professional individuals, who are appointed by the government, including the engineer, secretary to the commission, and his assistant. This board is presided over by the Minister of Public Works, and in his absence by the Under Secretary of State for that department.

This mixed commission, called the "Commission des Phares," is composed of naval officers, (of whom there is a majority,) of inspectors of the corps of bridges and roads, and of members of the Institute. It prepares the projéts for all new lights, and the general council of bridges and roads judges of the propriety of all schemes for that branch of service, under the four heads of architectural design, mode of executing the works, estimate of the expense, and the preparation of the specifications of the works. The light-house commission of France is not an administrative body, but is occupied solely in questions of principle or design, and leaves to the general directory of bridges and roads the care of providing the necessary means for the construction of new works, the expenses of illumination, &c.

The central commission at Paris is charged with the duty of providing all supplies necessary for keeping the illuminating apparatus in perfect order. There is also in Paris, belonging to this particular branch of the public service, a central workshop and depot, under the immediate care and supervision of the secretary-engineer to the commission, who superintends the construction (by mechanics employed by the administration) of all lanterns and their fixtures that may be required for the service; tests all apparatus before sending it to its destination; makes experiments upon all the optical and mechanical portions of apparatus destined for light-house purposes, combustibles, &c.; in short, this officer is charged with all the scientific details of the service, subject to the instructions, from time to time, which may be issued by the light-house commission. At this central depot are always kept, ready for immediate use, the various articles required in the illuminating department, such, for example, as mechanical and Argand lamps, glass chimneys, wicks, cleaning materials, &c.; also specimens of the different descriptions of apparatus used in light-houses, and apparatus constructed upon the latest and most approved plans ready for service.

All expenses incurred in the maintenance of the lights and their appendages are defrayed by the agents of the national treasury, from funds authorized by annual appropriations for those specific purposes.

No light dues are charged upon shipping in France as in Great Britain, Holland, Denmark, Norway, and Sweden, &c.; but the whole establishment is provided for as in the United States and Russia.

The maintenance of the light-house buildings is confided to the departmental or local engineers, and the expenses are defrayed from funds appropriated for the service of the department of public works.

The establishment of new works is decided upon by the Minister of Public Works, under the advice of the light-house commission. The determination of the minister is reported officially by the secretary of the commission to the Under Secretary of State for that department, and through his office to the prefect of the department in which the proposed work is to be established. The prefect directs the chief engineer of bridges and roads for that department to have detailed plans and estimates prepared upon the basis of the proposition of the light-house commission; these plans and estimates are transmitted through the office of the Under Secretary of State to the secretary of the light-house commission, who makes a report to accompany them to the light-house commission. The plans and estimates are then submitted to the light-house commission, which decides whether or not the wants of the service, *nautically or otherwise*, are such as to require the construction of the proposed works. In the preparation of these plans and estimates the military engineer of the department is consulted, to ascertain his opinions as to the propriety of constructing these works with reference to the defences of the coast.

The details having been completed, after having undergone the strictest scrutiny in every particular, the projét is presented to the general council of bridges and roads, to be considered with reference to the architectural designs, mode of construction, estimates of expense, &c. Having been approved by the general council of bridges and roads and the Minister of the Interior, the plan is then sent to the prefect of the department in which the light is to be established, with instructions to enter into contracts for the execution of the works, under the specifications and limitations authorized by the administration.

The execution of these works is entrusted to the engineers of bridges and roads for that department. As the works advance, the contractor receives payments upon the certificates of the engineers in charge, approved by the prefect of the department, from the departmental paymaster, (as deputy of the public treasury,) and the sums are charged to the budget for works of navigation, under the head of light-houses.

The light-house towers of France are constructed in the most substantial and perfect manner possible, without there being any appearance of unnecessary or wasteful expenditure. Great care is taken in the interior arrangements of the buildings, so that they may best answer the requirements of the service. Many of the towers are constructed of a soft stone of a rather peculiar kind, which hardens by exposure to the action of the atmosphere; those constructed of that material are lined inside with brick, leaving a sufficient space between the interior of the outer wall and the brick to allow a free circulation of air, thereby securing the building from dampness. Hard burnt bricks are preferred for light-house towers, when circumstances will admit of their being employed, particularly in fitting up the oil apartments, which are placed below the surface of the earth, to insure as equable a temperature during the whole year as may be possible to attain. The keeper's apartments are finished and fitted up in a plain, substantial, and economical manner, combining all the necessary accommodation and comfort. There is a room fitted up and properly furnished for the accommodation of the engineer, inspector, or other person authorized to make official visits, at each light station. Especial care is taken to secure proper ventilation to the towers and lanterns—all the necessary fixtures about the light-rooms, lanterns, apparatus, &c.--the most minute, and apparently unimportant details in the exterior and interior arrangements; in short, nothing could combine greater perfection in stability, in usefulness, and a proper economy, than is perceptible in everything connected with the light-houses visited by us on the coasts of France.

The repairs of the light-houses and their appendages are projected and executed by the engineers of the different departments in which they exist, who are limited as much as possible in their expenditures by the estimates of each year for those specific purposes. In some cases the contractor-general is authorized to make repairs, under the direction of the agents of the administration of bridges and roads.

Whenever application is made for a new harbor light, the subject is submitted to a local commission, assisted by the engineers of the department. The report is discussed by the light-house commission, and the same course subsequently followed as in the case of large or seacoat lights.

All the light-house towers in France are furnished with lightning conductors, made of copper wire twisted into the form of a rope, and about three-fourths of an inch in diameter.

In the organization of the lighting service, two systems are followed—the contract and the administrative. The ocean and Mediterranean coasts are under contract at present for nine years from 1839, for all the detail supplies of the service.   *   *   *   *
Among the clauses and conditions, it will be perceived that the contractor-general is required to be represented by a deputy in each department in which there are any lights; that the *oil of colza, clarified and refined*, must be used exclusively; and that the prices of oil will be regulated quarterly, based upon the average prices of the principal market in the kingdom for that particular article of commerce. M. Fresnel insists that this last clause has had a most salu-

tary effect of insuring the best oil the market could produce, without the contractor running any risk of loss. On the coast of the channel, from the frontier of Belgium to St. Malo, this service is performed by the administration, except for the article of oil, which is procured under a contract entered into for three years. That portion of the coasts of France which is lighted by contract includes even the salaries of the light-keepers; but where the service is performed by the administration, the keepers are appointed by the prefect of the department, upon the recommendation of the engineers. The smaller articles necessary to the illumination are sent from the central depot in Paris, under the charge of a conducting steward. The mechanical lamps are sent to Paris to be repaired under the engineer-secretary to the light-house commission. The administrative system recommends itself, for the reason that it avoids all intervention of interest foreign to that for which the lights were established. The contract system has been for a long time preferred in France, for reasons of economy, complication of accounts when performed by the administration, &c.; but the experience of the last seven years on the channel coast has sufficiently demonstrated the importance of changing it to the administrative; and it is deemed quite probable that, after the expiration of the present leases, that system will be exclusively adopted, except for supplies of oil.

The superintendence of the lights of France is confided to the local engineers of the corps of bridges and roads. The secretary to the light-house commission visits, each year, one of the three divisions into which the coast is divided, and his assistant another, so that the inspections, as far as possible, are biennial for each division. Monthly returns are made of all stores on hand, of the quantity of oil consumed each night, &c., to the secretary of the commission. These returns are intended as checks upon the keepers and answer the purpose admirably. A most rigid supervision is required at the hands of the inspecting engineers; and moreover, that they employ all possible means to detect any delinquency on the part of the keepers, or other agents connected with the service. It is conceded that all these precautions may fail to produce the desired effect, but that under such a supervision few among the guilty will escape detection. The lights visited by the undersigned were clean, and presented every indication of a perfect and systematic attendance and supervision.

Indications of the range of visibility afford very meagre data for forming a correct idea as to the relative value of apparatus for illumination. It is impossible to determine with certainty the *absolute range* of any light, in consequence of the different conditions of the atmosphere, and of the capacities of different observers. A first order dioptric light has been seen *fifty* miles very often, and one of the fourth order as far as sixteen miles. M. Fresnel says, upon the subject of range: "We would, then, draw very erroneous conclusions as to the relative value of the useful effect of the apparatus of these lights, in taking for a basis of comparison the indications of *range*, which are never fixed, or positive."

At the present time there are two systems of illumination in France—the old or reflector system, and the new or dioptric system. In 1822, M. A. Fresnel placed the first dioptric apparatus ever successfully employed, in the tower of Cordouan, at the mouth of the Gironde. In 1825, the light-house commission decided upon the exclusive use of the lenticular apparatus for the illuminations of the coasts of France and colonies; adopting, at the same time, the programme and report of Rear Admiral de Rossel, who had been charged, as a member of the "Commission des Phares," with that service.

Since that period new lights have been established, and old ones replaced with this new apparatus, until, on the 31st December, 1845, there were, of the two hundred and nine lights of every description belonging to the light-house department of France, one hundred and nineteen fitted with that apparatus. The remaining ninety lights were reflector lights, fitted with the Bordier Marcet (called "sideral") reflectors, and the parabolic reflectors, similar to those used in Great Britain and America. Of these last ninety lights, seventy-seven are small harbor or temporary lights, fitted, in most cases, with a single parabolic or Bordier Marcet reflector, marking the entrance to some channel or harbor. The remaining thirteen are fitted with lenticular apparatus of the most approved construction, in accordance with the original plan of 1825.

Engineers and other scientific and philanthropic individuals, of most if not of all the nations of the world, have made this new system of illumination an object of study and of critical examination; the results of which have been the successful though gradual application of it to the coasts of nearly all the commercial nations.

On the 31st December, 1845, *eighty-three* light-houses belonging to foreign governments had been fitted with lenticular apparatus constructed in Paris, to which may be added those constructed in England and Holland, say from fifteen to twenty, making, including those on the coasts of France, upwards of two hundred and ten; one hundred of which may be put down as of the three first orders, and the remaining one hundred and ten of the fourth order. These numbers do not include those at present in the course of construction for France, Egypt, (tower at Alexandria,) Brazil, and the colonies, islands in the Pacific, &c. M. Fresnel says, with perfect truth and reason: "After these numerous and extended applications, the dioptric system of lights may be fully appreciated under the double aspect of *theory* and *practice;* and I will add, that under the first point of view the question has been for a long time out of controversy."

There are six different orders of lenticular apparatus at present employed, viz: first, second, third larger model, third smaller model, fourth larger model, and fourth smaller model.

The different orders are subjected to different combinations, such as dioptric, two catadioptric—one with concave mirrors, and the other with catadioptric zones, or rings of glass, in triangular profile sections—and the "diacatopric,"* combining the dioptric portion and the catadioptric zones surmounted by plane mirrors. In addition, a spherically curved metallic reflector or mirror is placed on the land side of all lights which are only required to illuminate from four-fifths to five-sixths of the horizon, which reflects the rays from that side back through the opposite lenses.

"There can be no doubt," says a distinguished engineer,† who has had much to do with the light-houses of Europe, "that the more fully the system of Fresnel is understood the more certainly will it take the place of all other systems of illumination for light-houses, at least in those countries where this important branch of administration is conducted with the care and solicitude which it deserves." "To the Dutch belongs the honor of having first employed the system of Fresnel in their lights." "The commissioners of northern lights followed in the train of improvements, and in 1834 sent Mr. Alan Stevenson on a mission to Paris, with full powers to take such steps for acquiring a perfect knowledge of the dioptric system, and for forming an opinion of its merits as he should find necessary."

* See Mr. Alan Stevenson's report to commissioners of northern lights, for this word.
† Mr. Alan Stevenson, civil engineer.

"The singular liberality with which he was received by M. Leonor Fresnel, brother to the late illustrious inventor of the system, and his successor as secretary to the light-house commission of France, afforded Mr. A. Stevenson the means of acquiring such information and making such a report, on his return, as to induce the commissioners of northern lights to authorize him to remove the reflecting apparatus of the revolving light at Inchkeith, and substitute dioptric instruments in its place." * * * "The Trinity House followed next in adopting the improved system." * * * "Other countries begin to show symptoms of interest in this important change; and America, it is believed, is likely soon to adopt active measures for the improvement of their light-houses. "Fresnel, who is already classed with the greatest of those inventive minds which extend the boundaries of human knowledge, will thus, at the same time, receive a place amongst those benefactors of the species who have consecrated their genius to the common good of mankind; and wherever maritime intercourse prevails, the solid advantages which his labors have procured will be felt and acknowledged."

The fourth order lenticular lights are illuminated ordinarily by means of a common fountain, or constant level lamp and Argand burner, with a single cylindrical wick of three-fourths to seven-eighths of an inch in diameter, consuming about one and a quarter ounce of oil per hour, and forty-eight gallons per annum. The larger lights require mechanical lamps with multiple wicks, to as great a number as four, placed in concentric tubes, and the oil supplied to them by means of pumps, put in play by clock machinery. Hydraulic and pneumatic lamps have been employed in the place of the mechanical ones, but, with good reason, they are not approved of in France. For the catadioptric apparatus of half a metre in diameter, the ordinary constant level lamps, with two concentric wicks, burning about four and a half ounces of oil per hour, have been employed very successfully at several points on the coast of France, where the ordinary range of a light of the third order, for example, was not required, or for harbor lights requiring a powerful ray, or one whose brilliancy it is necessary to weaken by the application of a red chimney, with the view to give it a distinctive character. These double wick ordinary lamps require only one keeper to attend to them. Some of the burners in France are fitted with flat wicks for small and temporary lights, although by no means common, and generally disapproved of.

The dioptric lights of France are divided into six different orders ; but, with reference to their distinctive characteristics and appearances, this division does not apply, inasmuch as, in every order or class, lights of precisely the same character may be found, differing only in the distance at which they can be seen, and in the expense of their maintenance. The six different orders, as before mentioned, are not intended as distinctions, "but are characteristic of the power and range of lights, which render them suitable for different localities on the coasts, according to the distance at which they can be seen." "This division, therefore, is analagous to that which separates the lights of Great Britain into sea lights, secondary lights, and harbor lights—terms which are used to designate the power and position, and not the appearance of the lights to which they are applied."

In France there are nine principal combinations of lights possessing distinctive characteristics. These distinctions, for the most part, depend upon the periods of revolution rather than upon the characteristic appearance of the light. They are—

1. Flashes, which succeed each other every minute ;
2. Flashes, which succeed each other every half minute ;
3. Flashes, alternately red and white ;
4. Fixed lights, varied by flashes every four minutes ;
5. Fixed lights, varied by flashes every three minutes ;
6. Fixed lights, varied by flashes every two minutes ;
7. Fixed white lights, varied by red flashes more or less frequent;
8. Fixed lights ;
9. Double fixed lights.

There are very few double fixed lights in France. They are, however, sometimes employed for the purpose of giving a very decided character to the locality. For example, the first order lights at La Hève, near the port of Havre, and the two lights at present in the course of construction on the left bank of the *Canche*. Red fixed lights are not employed on the coasts of France, except as a distinguishing characteristic for harbor purposes. They are doubly objectionable : first, because of the great diminution of light in consequence of the absorption of the red glass chimney; and, secondly, it loses its distinctive character in foggy weather—all lights assuming a reddish tint under those circumstances.

The revolving reflector lights are objected to because of the fact that, ordinarily, they are only distinguishable by the duration of their eclipses, which often become positive at a very short distance from the light-house, and the interval of time between any two eclipses could not be extended to a greater limit than three minutes without prolonging the duration of the eclipses to such an extent of time as to mislead the navigator by depriving him for so long a time of his point of recognition. In the revolving dioptric apparatus, upon the latest and most approved plan, the duration of the eclipses is scarcely perceptible ; the fixed subsidiary parts of which reflect a light constantly visible in a horizon extending nine or ten nautical miles with a second order, and from twelve to fifteen with a first order apparatus.

The three first of the principal combinations only are applied to the first three orders, in consideration that in the inferior orders the flashes would have too short a duration, and the eclipses would be positive at too short a distance from the light, in consequence of the feebleness of the ray produced by the fixed subsidiary part of the apparatus.

The distinguished engineer, secretary to the "Commission des Phares," of France, M. Leonor Fresnel, kindly furnished the under-signed with the results of numerous photometric experiments which were made for the purpose of testing the comparative useful and economical effects of the two systems of illumination, to which they beg leave to call particular attention.

M. Fresnel says, in his note referred to, "the foregoing results confirm the following principles :

1. "The *useful effect* of a parabolic reflector increases with its dimensions and with that of the illuminating body.

2. "The *economical effect* of a reflector of given dimensions is greatest when the lamp-burner is smallest.

3. "The *divergence* is greatest when the flame is most voluminous, or when the reflector is smallest. We cannot, then, (all other things being equal) augment the *economical effect* of a reflector without diminishing its useful effect—that is to say, without reducing its brilliancy or intensity, and consequently its range (portée.)

"The reduction of the volume of light within certain limits is particularly objectionable when it appertains to eclipse apparatus, in which case it limits the width of the luminous cone, and consequently augments the length of the eclipses. The same reduction applied to the foci of reflectors composing a fixed light apparatus may weaken the light in their intervals to such a degree as to produce *dead angles*, or become completely obscured to the observer beyond certain distances.

"It is further proper to remark that the horizontal divergence is not lost for *useful effect*, but that the divergence, in the vertical sense, only profits the navigator in the limited angular space comprised between the tangent at the surface of the sea and the ray terminating at the distance of some miles from the light.

"Finally, there is for the calibre of the lamp-burners applicable to reflectors of given dimensions, and destined for the illumination of an equally determined range, *a maximum* beyond which prodigality of light ensues, and a *minimum* within which the illumination becomes insufficient."

The third order smaller size lenticular apparatus may be illuminated with very decided advantages by means of an ordinary Argand burner and single wick. Such a light would consume about two ounces of oil per hour, and is admirably adapted for harbor lights. In ordinary weather such a light may be seen from twelve to fifteen miles. One keeper alone can attend to all the duties of such a light, and it is maintained in France at an annual expense of about two hundred dollars.

M. Fresnel remarks, with reference to the ranges of different lights, their useful effect, &c. :

"The useful effect of a light-house apparatus is measured by the quantity of light which it projects upon the horizon. Observations of *range* for that purpose furnish very uncertain evidences, on account of the difficulty of ascertaining the *absolute range* of a light, which varies according to the state of the atmosphere and according to the good or bad sight of the observers."

Reflector lights, with not more than six or eight burners, are attended by one keeper, occasionally assisted by the members of his family. For lights with a larger number of burners, two keepers ; and if the light be in an isolated position, three keepers are allowed, with, in the latter cases, certain privileges not accorded to others.

Dioptric lights of the fourth order and third order smaller size, require but one keeper, except when in isolated positions. Two keepers are allowed to lights of the third order larger size, and for those of the second order, in consequence of the employment of the mechanical lamp.

First order lights are allowed three keepers ; and when there are two first order lights forming one combination, five keepers are allowed for the two lights. Lights of the first order in isolated positions, are allowed four keepers, and for the third order larger size and the second order lights, similarly situated, three keepers are allowed.*

In comparing the two systems of illumination, they should be considered under the heads—first, of absolute useful and economical effect; second, of first cost, repairs, and maintenance; and, third, of the facility and safety of the service.

The brilliancy of a catadioptric apparatus of 11.8 inches interior diameter, lighted by a lamp burning forty-five grammes of oil per hour, has been found, by photometric experiments, to be equal to eight or nine Carcel burners; while that of a "sideral" reflector of Bordier Marcet, illuminated by a lamp consuming fifty grammes of oil per hour, has been found, in the same manner, equal to only four burners of Carcel ; or, in other words, the brilliancy of the former is to the latter as one to two. The useful effect of the catadioptric apparatus, illuminating three-fourths of the horizon, is represented by 137,700, and that of the reflector by 68,400, which gives the value as one to two.

The economical effect of the catadioptric apparatus is represented by 3060, and that of the reflector by 1368 ; giving the value in that respect as 1 to 2.24.

No combination of reflectors can produce an equivalent to the third order smaller size apparatus, illuminated by an ordinary fountain lamp and Argand burner, with one wick, consuming sixty grammes of oil per hour, or one burner, with two wicks, consuming one hundred and fifteen grammes of oil per hour. An apparatus of this sort, with a lamp of two wicks, may be seen in ordinary weather (the horizon of the light, from its elevation above the sea level, being equal to or greater than that distance) at the distance of fifteen to eighteen nautical miles.

The brilliancy of a catadioptric third order larger size apparatus, illuminated by a mechanical lamp of two wicks, consuming one hundred and ninety grammes of oil per hour, (six and three-fourth ounces) has been found equal to seventy burners.

We suppose that it embraces only four-fifths of the horizon. To

---

* In England, Scotland, and Ireland, no difference is made between the number of keepers for dioptric and reflector lights.

illuminate, by means of reflectors, the same angular space of 288°, with an *effect of light about equal*, fourteen parabolic reflectors, of about eleven inches in diameter, illuminated by Argand lamps, consuming each thirty-five grammes of oil per hour, will be required. The useful effect of these reflectors will be represented by 870,240, and that of the catadioptric apparatus by 1,160,000 ; and thus it is seen, that notwithstanding the very great difference in favor of the catadioptric apparatus, in the consumption of oil, it is also superior in useful effect to the light with the fourteen parabolic reflectors. Further, the economical effect of the catadioptric apparatus is represented by 6105, and that of the reflector by 1776, or as 1 to 3.44 : "that is to say, without estimating the expenditure of oil by *unity of light*, the lenticular light will be nearly three and a half times more advantageous than the reflector light." With regard to the effective expenditure of oil, they will be in the proportion of 190 grammes to 14 × 35 grammes per hour, or as 1 to 2.6.

The brilliancy of a catadioptric apparatus of the second order, with a mechanical lamp of three concentric wicks, consuming five hundred grammes of oil per hour, has been found equal to two hundred and sixty-four burners. Supposing that it is only required to illuminate three-fourths of the horizon ; then, to obtain an effect about equal in angular space of 270°, at least thirty-four parabolic reflectors of about twenty inches in diameter will be required, which will give a useful effect which is represented by 3,525,120, while that of the catadioptric apparatus is represented by 4,120,000. The comparison between the absolute consumption of oil will be equal to 2.86 to 1 ; and that of the quantity of oil expended by unity of light equal to 3.33 to 1 : thus, under this last report, the lenticular apparatus will be three and a third times as advantageous as the catoptric apparatus.

The maximum brilliancy of a revolving light of the second order, with twelve lenses, has been found to be equal to 1184 burners, and its minimum brilliancy equal to one hundred and four burners. To construct a light, with parabolic reflectors, possessing an equal effect, it will require twenty-four with diameters from twenty-two inches to twenty-four inches, arranged on six faces of the revolving frame. In making the comparison, however, for want of precise data as to the lustres of those reflectors, those of about twenty inches diameter will be referred to. It is supposed that the two lights compared are constructed so as to present the same distinguishing features ; the maximum lustre of the reflector light will be equal only to 1080 burners, with other disadvantages; for the details of which, reference may be made to M. Fresnel's note No. 1, section two, (hereto annexed.) M. Fresnel remarks, in this connection, "Without pressing further the comparison of the effects of the two kinds of apparatus, we will perceive, without doubt, the evident advantages of the dioptric or lenticular combination, which in fine weather will not present *an absolute eclipse* at a less distance than from fifteen to eighteen nautical miles. If we now consider the expenditures of oil, we will find, first, that they are as 24×42 is to 500, or as 1 to 2; second, that the economical effects will be as 2469 is to 10,043, or as 1 to 4.07 : thus the lenticular apparatus will be four times as advantageous as the reflector apparatus." "Let us remark, before proceeding further, that in employing 24 parabolic reflectors, of about 20 inches diameter, for such an apparatus, we reach the utmost possible' limit, without admitting the employment of lanterns of a size beyond all proper bounds; and we may also affirm, that very few of the catoptric lights, considered as lights of the *first order*, equal the lenticular lights of the same character of the second order.

With reference to the first order dioptric lights, M. Fresnel remarks, in his note : "Now, we have found that the total lustre or brilliancy of an apparatus of this kind is equal in all its azimuths to 480 burners of Carcel. But it will be practically impossible to obtain a like effect in the catoptric system, without having recourse to the employment of 36 parabolic reflectors of about 24 inches diameter." "The difficulty becomes still greater, if it be necessary to attain with these reflectors the effect of a revolving lenticular light, with eight large lenses, the lustres or flashes of which exceed 4,000 burners of the Carcel lamp."

"Let us limit ourselves, then, without entering into more full details, to the observation, that the *economical effect* of a fixed light of the first order, illuminating three-fourths of the horizon, is to the *economical effect* of a light composed of parabolic reflectors of about twenty-inches diameter, as 10,080 to 2,469, or as 4.08 to 1 : that is to say, that the first will be (as to the expense of the oil only) four times as advantageous as the second."

With regard to lights *varied by flashes* or *short eclipse lights*, "the catoptric system is not susceptible of producing that combination without great difficulty, which unites to the permanence of fixed lights the advantage of presenting a very decided character." "No repairs are required upon the lenticular apparatus.   *   *

The amount necessary to construct and put into operation a "sideral" light for harbor purposes may be stated at 8,150 francs, or about $1,500; and the annual expense for its maintenance, including interest upon the cost at the rate of five per cent., at 1,207 francs, or about $225.

The amount necessary for a catadioptric smaller model harbor light may be put down at 9,181 francs, or about $1,700 ; and the annual expense for maintenance, including interest of first cost, &c., as above, at 1,259 francs, or about $235.

The useful effect of the "sideral" light has been found equal to 68,400, and its economical effect represented by 57.

The useful effect of the catadioptric light, illuminating three-fourths of the horizon, has been found equal to 137,700, and its economical effect, after the same manner, is represented by 109. The comparison of these two will, then, be in the proportion of 57 to 109, or as 1 to 1.91. "Then, besides the advantages of *a double lustre,* the catadioptric apparatus, in an economical point of view, is nearly twice as advantageous as the catoptric apparatus."

M. Fresnel remarks : "It is difficult to establish a comparison of a precise kind between the fixed lights of the third order in the old and the new systems, because we cannot obtain with the ordinary parabolic reflectors a passably equal distribution of light, without multiplying those reflectors to such a number as would require a much greater expenditure of oil than could be allowed for lights of that class." He says further : "I will merely observe that I have every reason to believe, from the indications contained in the table of light-houses of the United States, that among all the lights of that country illuminated by reflectors, the diameters of which do not exceed sixteen English inches, there are very few whose *useful effect* is superior or equal to that of a catadioptric light of the third order larger model."

The amount necessary for establishing a reflecting revolving light with twenty-four parabolic reflectors of about twenty inches diameter, is estimated at 73,000 francs, or about $13,700.

Annual expense for maintenance of the same, including interest at five per cent. per annum, will be 8,650 francs, or about $1,625.

The amount necessary for establishing a second order revolving lenticular light is estimated at 105,500 francs, or about $19,800.

The annual expense for maintenance of the same, including interest at five per cent. per annum, will be 11,075 francs, or about $2,075.

The useful effect of the reflector light is represented by 2,488,320, and its economical effect by 288.

The useful effect of the lenticular light is represented by 5,021,467, and its economical effect by 453.

The economical effect of these two lights will then be represented by 288 and 453, or in the proportion of 1 to 1.6. "From whence it results definitively that the lenticular light of the second order will be more than *one and a half times* as advantageous as the catoptric or reflector light, which we may without doubt consider as being of the first order, and the *useful effect* of which, nevertheless, could not be equal to *but half* of the *useful effect* of the former."

No comparison can be entered into between the first order lenticular lights and reflector lights, for the reason that it is impossible to construct a reflector light which would produce a sufficiently powerful effect to be compared to a dioptric one, without increasing the dimensions of the lantern, and the number and size of the reflectors, to a degree which would be attended with a very great expense, and equally great inconvenience.

From the foregoing details, which have been drawn mainly from information furnished by M. Fresnel, the following seems to be but just conclusions :

"1. That the lights fitted with the dioptric apparatus present a variety in their power and effects, and may be made to produce an intensity of lustre, which render them of an interest, in a nautical point of view, incontestably superior to those fitted with the catoptric apparatus.

"2. That if we take into account the first cost of construction and the expense of their maintenance, we will find, with respect to the effect produced, the new system (dioptric) is still from *once and a half* to *twice* as advantageous as the old," (reflector.)

If additional arguments and evidence were wanting to establish the now almost universally conceded fact, of the very positive and decided advantages of the dioptric system of Fresnel over all other modes of illumination for light-houses, they might be found to exist at present in an unanswerable form—that of the practical and successful application of the system, within the last few years, in nearly all the commercial nations of the world. Prior to the year 1832, there was not a single dioptric light out of France; and on the French coast, at as late a period as 1834, there were but fourteen large and fifteen small, or harbor lights, fitted with the dioptric apparatus.

On the 31st December, 1845, there were belonging to the French light-house department *one hundred and twelve lights* fitted with the dioptric apparatus, and throughout the world not less than *two hundred and ten lights* fitted upon this new system ; one hundred of which are of the three first orders, and the remaining one hundred and ten, small or harbor lights, without including apparatus now in course of construction at Paris, to which allusion has already been made.

The objections which have been made by *a few persons* to the employment of the Fresnel dioptric apparatus for the illumination of light-houses, in consequence, as they allege, of the difficulties which attend the management of the mechanical lamps with concentric wicks, (which are absolutely necessary for the proper illumination of the larger orders of apparatus,) seem to be no longer tenable, if indeed there ever were any reasonable grounds of objection on that account.

The twenty-three years' experience in France, (dating from the time the Cordouan light was exhibited,) where *ordinary day laborers are taken for light-keepers*, and the undeniable fact of the successful employment of the system for fourteen years in Holland, Scotland and Norway ; for from five to ten years in England, Sweden, Denmark, Prussia, Belgium, Spain, Sardinia, Tuscany, Naples, Brazils, West Indies, islands of the Pacific ocean, Cape of Good Hope, &c., must be sufficient evidence to convince any disinterested and unprejudiced mind of the utter folly of such an objection at the present day.

In a communication to the government of Norway and Sweden, in 1830, M. Fresnel remarks upon this subject : "Happily, an experience of *seven years* has dissipated that fear, and the lenticular lights have been distinguished up to this time by the regularity of their service." Again, in reference to the same subject, M. Fresnel remarks, in a note to the undersigned, that "opinions thus expressed *fifteen years since*, based upon an experience of seven years, have been greatly strengthened up to the present time, embracing a period of *twenty-two years* since the establishment of the Cordouan light, and sustained by the results daily offered of more than one hundred and ten lights of the first three orders, established along the coasts of France and different foreign powers." "In this important point of view, then, the question seems to be irrevocably settled; and I will only add a few considerations relative to the application, more or less extended, which may be made of the new system of illumination to the vast maritime coasts of the United States."

It has been further objected, that competent persons could not be procured in the United States to take charge of the lights fitted with the dioptric apparatus and mechanical lamps, for the salaries at present paid to light-keepers of the existing lights. The number of keepers necessary for those lights has also been urged as an objection to their introduction; and there is also a third objection, emanating from the same source, that the mechanical lamps could not be repaired when employed at distant or isolated points on the coast.

With regard to the keepers, no better evidence can be adduced than the opinions of M. Fresnel upon the subject, and the practical results furnished daily wherever the lights are employed. M. Fresnel says, "that the difficulty of obtaining proper persons to fill these subaltern stations appears to be most singularly exaggerated." "In France they belong almost always to the class of *ordinary mechanics or laborers*, who make from one and a half to two and a half francs per day, (from 27 to 46 cents.") "Eight or ten days will suffice, ordinarily, to instruct a light-house keeper in the most essential parts of his duty, receiving lessons from an instructor conversant with all the details of the service; and two instructing officers will be sufficient to prepare keepers for all the lenticular lights which could be successively established upon the coasts of North America." "In defence of this assertion, I will cite the example of the administration of Norway and Sweden."

As to the number of keepers allowed to the dioptric lights, there might be some reason in the objection, if it were possible to produce a light with parabolic reflectors possessing in any reasonable degree the advantages arising from the employment of a first order catadioptric apparatus; but as it is well established that reflectors are not susceptible (practically) of any combination which would produce a light equal in every respect to a first order dioptric light, the objection ought in honesty to be abandoned or waived by them, without they prefer bad to good lights, to guide the mariner in his perilous way along our shores.

The lower orders of dioptric apparatus, illuminated by ordinary Argand lamps and burners, with single and double wicks, require but

one keeper ; and they produce a light far superior to those of the
same class in the catoptric system, independently of the economy in
the use of the dioptric lights.   In Scotland and in England, where
the lights are as well if not better attended than in any other parts
of the world, the same number of keepers are allowed for the same
class of lights, without regard to the apparatus employed, whether
catoptric or dioptric.   At the South Foreland, for example, there are
only three keepers for a first order dioptric and a first order reflector
light, placed about three hundred yards apart ; and at St. Catherine's
a first order dioptric light has but two keepers to attend it ; besides,
other instances might be cited, if it were deemed at all necessary.
But to accomplish in the most perfect manner possible the great and
important objects for which lights are established upon seacoasts, it
would seem but reasonable, and certainly desirable, rather to increase
the number of keepers ordinarily allowed to catoptric lights, than to
diminish the number (taking France as a basis) for those fitted with
dioptric apparatus.

In regard to the repairing of the mechanical lamps, it may be asserted, without the fear of being controverted, that in consequence
of the superior manner in which these lamps are at present constructed in Paris, they will perform well for a number of years by bestowing upon them only the ordinary attention necessary to keep them
clean ; besides, the number supplied to each light-house (from 3 to
4, and never less than 3,) is a sufficient guarantee against any accidents which could prevent the proper exhibition of the lights.   The
same objections might, with equal propriety, be urged against revolving, flashing, or any other lights requiring clock machinery ; yet
such lights are found on every coast where lights exist to any extent.
A simple inspection of the works of a mechanical lamp will convince
any person, of common understanding, that any mechanic who is capable of repairing the machinery for a revolving light is equally competent to put in order any lamp used in light-houses, and particularly
those known as mechanical lamps with concentric wicks.

The oil of colza is used exclusively in the French light-houses.   M.
Fresnel says : "From numerous experiments, it seems to me that these
two oils (spermaceti and colza) may be employed with equal success
in lamps of single or multiple wicks."

M. Fresnel's preference for the colza (to the sperm oil) is based
upon two reasons : first, the colza is less expensive in France than
sperm, owing to the fact that the vegetable from which this oil is

expressed, is cultivated on a very extended scale in France, Belgium, Holland, Holstein, &c. ; and, secondly, the great difficulty in detecting impositions which may be and are practised by mixing inferior oils with the sperm, while, on the other hand, any impurities in the colza are very readily detected. No experiments have yet been made in France to test fully which of the two kinds of oil will produce the best light for light-house purposes.

*   *   *   *   *   *   *   *

There is but one floating-light in France ; that is constructed of wood, moored and illuminated after the manner, with a few exceptions, of those belonging to the Trinity Board in England. The exceptions are—first, bronze is used in the construction of the lantern in the place of iron ; and, secondly, the lamps are mechanical, the pumps of which are put in play by springs, instead of the ordinary fountain lamp. This latter, in spite of the delicate machinery of the lamp, is deemed a very decided improvement, as fulfilling much more fully the requirements of such a lamp, by preserving the centre of gravity in the same vertical during the whole time of the combustion.

HOLLAND.

Ventilation tubes are used in all the lanterns to conduct off the smoke and gasses from the burners. There are ventilators in the floor and sides of the lanterns, which are opened and shut as circumstances require.

The general opinion is, that the dioptric system of Fresnel is preferable to the catoptric one for the reasons — first, the superior quality and intensity of the light; second, economy in the maintenance when a great arc of or the whole horizon is required to be illuminated ; third, the variation which can be given to the revolving lights to distinguish the one from the other, by means of determined eclipses, either short or long, or alternated with flashes, and remaining at the same time fixed, visible at shorter distances.

The large number of officers who are constantly employed by the Netherlands government in examining the condition of the dikes, &c., on the seacoasts, affords ample means for discovering any want of efficiency in the lights, or attention to the duties devolving upon those in charge of them.

DANISH LIGHT-HOUSES.

*   *   *   *   *   *   *   *

A due regard is had to the ventilation of the lanterns, to prevent smoke, insure a proper combustion of the oil, and a pure atmosphere for the keepers to breathe, which latter has been of the greatest importance in preserving the health of those whose duty requires them to spend several hours of each night in an atmosphere which, without that precaution, must become injurious to health, to say nothing of other ill effects, such as producing an inferior light, &c.

*   *   *   *   *   *   *   *

Both the catoptric and dioptric systems are in use at present in Denmark. Prior to 1842, the catoptric system only was used ; but since that period *seven dioptric and catadioptric lights* have been established, which are highly approved of.

The lenticular apparatus were all made in Paris by Mr. Henry Lepaute, under the inspection of M. Fresnel, the Engineer-inspector, Secretary to the Commission des Phares of France, and put up in their respective towers by native Danes, who had had the opportunity of seeing similar apparatus put up in Norwegian light-houses.

NORWEGIAN, SWEDISH, AND RUSSIAN LIGHT-HOUSE ESTABLISHMENTS.

*   *   *   *   *   *   *   *

In 1830, the government of Norway and Sweden addressed a communication to the French light-house commission upon the subject of the improvement of its lights. In 1832, the light-house at the Isle of Oxöe was fitted with a dioptric apparatus, the manufacture of Monsieur Soleil, senior, of Paris, under the inspection of M. Leonor Fresnel.

The next dioptric apparatus for that coast was furnished by the same distinguished French artist in 1836, and placed in the tower at Gunarsborg. Since that period, dioptric apparatus have been annually introduced into the light-houses of Norway ; and, at the present time, there are not less than fourteen or fifteen of the different orders. These lights were fitted up by French mechanics, and the Norwegian keepers who were instructed in the management of the French mechanical lamp with concentric wicks. All lights of this description are now fitted up and attended to, exclusively, *by natives of the country;* and it has been reported that no difficulties have been encountered in any of the details relating to their management which required foreign aid.

The rape-seed or colza oil is employed in these lights. The usual means are employed for keeping the oil in a limped state during the cold weather, (stoves and frost lamps.)

PRUSSIAN LIGHT-HOUSES.

Of the eleven Prussian light-houses at present in operation, two only are fitted with the lenticular apparatus of Fresnel of the second order, which were constructed in Paris. The nine remaining lights are fitted with reflectors. Sufficient time has not elapsed since the introduction of the lens lights for the authorities charged with their management to form an opinion as to their superiority to the reflectors ; yet they are satisfied, from the reports of those familiar with them, that they are greatly superior to any other apparatus for light-house purposes.

*   *   *   *   *   *   *   *

CHARACTER OF THE FRENCH LIGHTS.

"Next to the necessity for visibility at a greater or lesser distance, lights ought, as far as possible, to present characteristic appearances by which to distinguish them not only from the fixed lights which may be visible in the same horizon, but also from lights situated within the limits of any reasonable errors that might be committed by navigators. That announcement would seem to render it necessary to discard the *fixed white lights.* However, as the revolving lights allow of but a small number of sufficiently marked characteristics ; and, moreover, as the fixed lights are ordinarily sufficiently distinguished by their elevation; and their greater brilliancy of light than those serving for the illumination of the interior of dwellings and of harbors, we have introduced this description of light, in a greater or smaller proportion, into the whole system of lighting the sea-coasts."

"We sometimes have recourse to *double fixed lights,* for the purpose of giving to them a very decided character ; for example, the two lights of La Hêve, near the port of Havre, and the two large lights which are in the course of construction at this time on the left bank of the Canche, in the department of the Pas de Calais, without speaking of the double small lights which serve to mark the direction of the channels leading to ports or roadsteads.''

"*Fixed red lights* (colored by the application of red glasses) are doubly objectionable: first, because of the very great absorption of the rays of light emitted from the focus; and, secondly, they afford only a doubtful character to the light during the existence of *fogs, when all lights assume a reddish tint.* This mode of distinction has not been employed on the coast of France, except as accessory, and then only for some small lights situated at the entrances of harbors, for the purpose of varying the appearance of some changing light.''

"*Revolving catoptric, or reflecting lights,* are distinguishable generally only by the greater or lesser duration of their *eclipses,* which become *positive* often at the distance of two or three nautical miles.''

"This mode of distinction, which would seem at the first instant to offer a great variety of combinations, is in reality a very limited one. On one side we cannot extend the interval of time between two consecutive eclipses to three minutes, without prolonging beyond proper limits the duration of the eclipse, during which time the navigator finds himself deprived of his point of recognition ; and, on another side, experience proves to us that the difference of thirty seconds may frequently escape an inattentive observer, or one who, in boisterous weather, becomes alarmed at his supposed dangerous situation.''

"The inconvenience arising from the duration of the eclipses is scarcely perceptible with the revolving dioptric apparatus upon the new model, the fixed accessory parts of which reflect a light constantly visible in a horizon extending to the distance of nine or ten nautical miles with an apparatus of the second order, and with those of the first order to twelve or fifteen nautical miles.''

"The different combinations of changeable dioptric lights at present established, are as follows :

First, flashes which succeed each other every minute.
Secondly, flashes which succeed each other every half minute.
Thirdly, alternate white and red flashes.
Fourthly, fixed light, varied by flashes once in every four minutes.
Fifthly, fixed light, varied by flashes once in every three minutes.
Sixthly, fixed light, varied by flashes once in every two minutes.

Seventhly, fixed white light, varied by red flashes more or less frequent."

"These make a total of nine principal combinations. It is proper to remark, besides, that the first three combinations have been applied only to the first three orders of lights, conceiving that in the inferior orders the flashes would have too short a duration; and that the eclipses do not cease to be positive, but at too short a distance from the light-house, in consequence of the feebleness of the fixed subsidiary portion of the apparatus."

USEFUL AND ECONOMICAL EFFECTS OF THE DIFFERENT LIGHTS.

"The useful effect of a light-house apparatus is measured by the quantity of light which it projects upon the horizon.

Observations of *range* for that purpose furnish but very uncertain evidence, on account of the difficulty of ascertaining the *absolute range* of a light, which varies according to the state of the atmosphere, and according to the good or bad sight of the observer.

To arrive at, and to establish, in this connection, true parallel valuations, it is indispensable to have recourse to photometric experiments. For that purpose, having taken *a model lamp for unity of light,* (the ordinary lamp of Carcel, for example,) receive upon a screen fixed at a proper distance the two shadows projected by the same *style,* which is illuminated at the time by the apparatus to be measured, and by the *lamp of unity.* The latter, placed upon a movable stand or table, is moved nearer to or further from the screen, until the two adjacent shadows appear of an equal intensity, and the squares of the distances of the two luminous bodies to the screen will be to each other as the intensities of their lights.

To measure the *total useful effect* of a parabolic reflector, or of a plano-convex lens, we may proceed in the following manner: After having placed the apparatus in the centre of a revolving stand or table, fitted with a movable needle and a graduated circle, measure the intensity of the light in different azimuths upon the whole horizontal extent of the luminous cone; take the mean corresponding to each angular extent; multiply that mean by the number of divisions of the arc, and the sum of all these partial products will represent the *total useful effect.* The *mean brilliancy* of the illuminated sector answers, moreover, to the quotient of the useful effect, divided by the arc which measures the amplitude of that sector.

We will find, lastly, the quantity of oil expended *by unity of light, and per hour*, in dividing the total useful effect by the number of grammes of oil consumed in an hour, and the quotient will represent *the economical effect.*[*]

Photometric measures taken in this manner in the whole illuminated angular extent of a revolving apparatus, make known the *maxima and minima* intensity of its flashes, and from which we determine the *duration of their apparitions*, by dividing by the entire circumference the product of the illuminated arc, multiplied by the duration of a revolution.

If it be required to construct with parabolic reflectors an apparatus for a fixed light to illuminate *a given angular extent*, we must first determine, by photometric results, the number of reflectors necessary to distribute in the extent to be illuminated—if not uniformly, at least in a complete manner.

Finally, if the catoptric apparatus is to be revolving, it is easily ascertained what number of reflectors are to be placed in the same plane upon each face of the movable frame, to obtain the total brilliancy necessary."

Useful and economical effect of the catoptric apparatus.

"Catoptric lights are so varied in their dimensions, and their reflecting powers at the same time so variable, according to the degree of perfection in their fabrication, (without adverting to the differences in the calibres of the lamp burners,) that it will be very difficult to present, with any degree of certainty, the table of *useful effects* to the principal lights of that system at present in use on the coasts of England, the United States, and of France.

I will limit myself here to some indications relative to lights of this description, which I have at my disposition. I will remark, once for all, that the *unity of light* to which all the photometric results which may be given hereafter have been referred, is the ordinary burner of a Carcel lamp, burning 42 grammes (1 oz. 7.73 drachms) of the oil of colza per hour."

---

[*] Whenever the *economical effect* of two lights is to be compared, it is not sufficient to take simply the expenses of the oil. It will be necessary to estimate their annual expense, and divide it by the amount of light usefully expanded upon the horizon, in the manner hereafter described.

"(A.)—Photophore (reflector) with an aperture of 50 centimetres, (19.69 inches,) and a depth of 20 centimetres, (7.876 inches,) illuminated by a lamp burning 42 grammes (1 oz. 7.73 drachms avoirdupois) per hour of oil.

| AZIMUTHS. | BRILLIANCY OR LUSTRE. | | OBSERVATIONS |
	Corresponding to the divisions.	MEANS.	
	In Burners.	*In Burners.*	The brilliancies or lustres measured at equal distances to the right and to the left of the axis have presented differences sufficiently remarkable; but uniformity has been established by taking the means. No account has been taken of the brilliancy of the simple burner beyond 10°; so that the amplitude of the luminous cone has been computed as reduced to 20°.
10°	2	10	
8°	18	31	
6°	44	59	
4°	74	117	
2°	160	215	
0°	270	215	
2°	160	117	
4°	74	59	
6°	44	31	
8°	18	10	
10°	2		
		864	This sum represents the *useful effect* for the uniform divisions of every 20°.

Useful effect, corresponding (as is ordinarily done) to the division of the circle into minutes, in this case$=864$ burners$\times 120'=103,680.$

Mean lustre or brilliancy$=\dfrac{864}{10}=86.4$ burners.

Maximum lustre or brilliancy corresponding to the axis$=270$ burners.

Economical effect$=\dfrac{103,680}{42 \text{ grammes}}=2,469.$

(B.)—Photophore (reflector) with an aperture of 0.275 metre, (10.83 inches,) and a depth of 12 centimetres, (4.72 inches,) illuminated by a lamp burning 35 grammes (1 oz. 3.75 drachms) of oil per hour.

AZIMUTHS.	BRILLIANCY OR LUSTRE.		ANGULAR DISTANCE.	PRODUCTS.	OBSERVATIONS.
	Corresponding to the divisions.	MEANS.			
	Burners.	Burners.			
16°	2	3	1°	3	As the divisions are un-
15°	4	5	1°	5	equal, it has been found
14°	6	9	2°	18	necessary to multiply each
12°	12	15	2°	30	mean lustre or brilliancy
10°	18	22. 5	2°	45	by the angular space cor-
8°	27	37. 5	2°	75	responding to it.
6°	48	57	6°	342	
0°	66				
				518	

Useful effect, corresponding to the division in minutes, in this instance, is $= 2\times518\times60'=62,160$.

Mean lustre or brilliancy$=\dfrac{518}{16}=32.4$ burners.

Maximum lustre or brilliancy corresponding to the axis$=66$ burners.

Economical effect$=\dfrac{62,160}{35\text{ grammes}}=1,776$.

(C.)—"Sideral" reflector of Bordier Marcet, formed of two parabolic metallic mirrors, with a diameter of 0.344 metre, (13.54 inches,) and an aperture of 0.187 metre, (7.364 inches,) illuminated by a lamp burning 50 grammes (1 oz. 12.25 drachms avoirdupois) of oil per hour.

Uniform brilliancy or lustre in all the azimuths$=4$ burners. The extent illuminated, after deducting the spaces occupied by the lamp and the frame, is about 285 degrees, or 17,100 minutes.

Useful effect$=4\times17,100=68,400$.

Economical effect$=\dfrac{68,400}{50}=1,368$.

The foregoing results confirm the following principles, the evidence of which springs moreover from a simple enunciation of them:

1st. *The useful effect* of a parabolic reflector increases with its dimensions and with that of the illuminating body.

2d. *The economical effect* of a reflector of given dimensions is greatest when the lamp-burner is smallest.

3d. *The divergence* is greatest when the flame is most voluminous, or when the reflector is smallest. We cannot, then, (all other things being equal,) augment the *economical effect* of a reflector, without diminishing its *useful effect*—that is to say, without reducing its brilliancy or intensity, and consequently its range. The reduction of the volume of light within certain limits is particularly objectionable when it appertains to eclipse apparatus, in which case it limits the width of the luminous cone, and consequently augments the length of the eclipses.* The same reduction applied to the crowns (foci) of reflectors composing a fixed light apparatus, may weaken the light in their intervals to such a degree as to produce *dead angles*, or become completely obscured to the observer beyond certain distances.

It is further proper to remark that the horizontal divergence is not lost for *useful effect*, but that the divergence in the vertical sense only profits the navigator in the limited angular space comprised between the tangent at the surface of the sea and the ray terminating at the distance of some miles from the light.

Finally, there is for the calibre of the lamp-burners applicable to reflectors of given dimensions, and destined for the illumination of an equally determined range, *a maximum* beyond which prodigality of light ensues, and a *minimum* within which the illumination becomes insufficient.''

Useful and economical effect of dioptric apparatus.

''Before commencing an analysis of the effect of the light of the lenticular or dioptric apparatus, I will state—

First. That these lenses always contain, independently of the fixed or movable dioptric drum, *a subsidiary catoptric* or *catadioptric part*, the *useful effect* of which is combined with that of the *principal part*.

* It was to obviate the defect of divergence that Bordier Marcet invented his large double parabolic reflectors with double focus; but that combination caused too great a loss of light.

Second. That when these lenses are not required to illuminate the whole of the horizon, that side which remains deprived of light is fitted with spherical reflectors (metallic) in place of the dioptric drum, which adds about *one-fifth* to the intensity of the sector directly corresponding.

That being stated, I now proceed to present the photometric results obtained from different lenses of the four orders of dioptric apparatus.

[Lenticular apparatus of the first order.]

(*A.*)—*First order lenticular apparatus for a fixed light, with accessory catoptric part.**

Lustre or *brilliancy* of the dioptric drum $=360$ burners.
Mean brilliancy of mirrors—

$$\left.\begin{array}{l} \text{7 upper zones} = 80 \\ \text{4 lower zones} = 40 \end{array}\right\} = 120 \text{ burners.}$$

Total amount of brilliancy $=480$ "

Calculation of the useful effect.

1st. The dioptric drum illuminating equally the whole of the horizon, (except about $26°$, being computed the amount of space occupied by the frames of the apparatus, and that of the lantern,)$=360$ burners $\times 20,000' = \ldots\ldots\ldots\ldots\ldots\ldots\ldots\ldots\ldots\ldots$ 7,200,000

2d. 7 zones of upper mirrors $=80$ burners $\times 20,000'=\cdots$ 1,600,000

3d. 4 zones of lower mirrors, omitted for one-eighth of the circumference, for the necessary passage of the keepers, $=40$ burners $\times 17,500'=\cdots\cdots\cdots\cdots$ 700,000

Total of useful effect $\cdots\cdots\cdots\cdots\cdots\cdots$ 9,500,000

Economical effect $=\dfrac{9,500,000}{750}$ grammes $= 12,667.$

º Notwithstanding the decided advantages which the catadioptric zones present, compared to the mirrors, we may, by reason of their economy, prefer, under some circumstances, the second combination to the first for lenticular apparatus of the *first order.*

If the apparatus illuminate only three-fourths of the circumference of the horizon, the useful effect will be reduced to$\cdots\cdots\cdots\cdots\cdots\cdots\cdots\cdots\cdots\cdots\cdots\cdots$ 7,200,000

To which must be added the effect of the reflector occupying the vacant space in the dioptric drum, one-fifth (360 burners×5,000′=$\cdots\cdots\cdots\cdots\cdots\cdots\cdots\cdots$ 360,000

Useful effect of the apparatus, for 270°=$\cdots\cdots\cdots\cdots$ 7,560,000

1st order, economical effect$=\dfrac{7,560,000}{750}$ grammes=10,080

(*a*¹)—*First order lenticular apparatus for a fixed light, with accessory catadioptric part.*

Lustre or *brilliancy* of the dioptric drum$\cdots\cdots\cdots\cdots$360 burners.

Cupola or upper zones=.$\cdots\cdots\cdots\cdots$140 burners $\Big\}$ 200   "
Lower zones=$\cdots\cdots\cdots\cdots\cdots\cdots\cdots$ 60   "

Total brilliancy$\cdots\cdots\cdots\cdots\cdots\cdots\cdots$ 560   "

Calculation of the useful effect.

560 burners×20,000′=$\cdots\cdots\cdots\cdots\cdots\cdots\cdots$ 11,200,000

From which must be deducted for the passage of the keepers, 60 burners×2,500′=.$\cdots\cdots\cdots\cdots$ 150,000

Remainder of useful effect=$\cdots\cdots\cdots\cdots\cdots\cdots$ 11,050,000

Economical effect$=\dfrac{11,050,000}{750}$ grammes = 14,733.

If the apparatus illuminate only three-fourths of the circumference of the horizon, the useful effect becomes reduced to 560 burners×15,000′= $\cdots\cdots\cdots\cdots$ 8,400,000

To which must be added, as in the other case, for the advantages of the reflector occupying the open space in the dioptric drum, one-fifth (360 burners×5,000′)= 360,000

Useful effect of the apparatus, for 270°$\cdots\cdots\cdots\cdots$ 8,760,000

Economical effect$=\dfrac{8,760,000}{720}$ grammes= 11,680.

(a²)—First order revolving lenticular apparatus, with accessory catoptric part.

Brilliancy or lustres, measured upon one-half of the amplitude of the luminous beam, emitted from a lens of one metre (39.38 inches) in height, and occupying a space of 45°:

Azimuths.	Brilliancy or Lustre.		Observations.
	Answering to the divisions	Means.	
	BURNERS.	BURNERS.	
210′	4 (?)	77	The annexed brilliancies of light are the results
180′	150	475	of means of different experiments.
150′	800	1,400	It is just to remark that all the divisions have
120′	2,000	2,450	intervals of thirty minutes, so as to distribute,
90′	2,900	3,300	in effect, the partial products of the mean
60′	3,700	3,800	intensities into the angular distances.
30′	3,960	4,050	
0′	4,200		
		15,552	

The *useful effect* of a lens $= 2 \times 15,552 \times 30′ = 933,120$

Ditto······of the 8 lenses of a dioptric drum ····· *7,464,960

Ditto······of 11 zones of mirrors, as before ······· 2,300,000

Total useful effect··································· 9,764,960

Economical effect $= \dfrac{9,764,960}{750}$ grammes $= 13,020.$

If the apparatus illuminate only three-fourths of the horizon, the useful effect will be reduced to 7,758,720; and the economical effect to 10,345.

○ We have given 7,200,000 to the dioptric drum of a *fixed light*—an amount which seems satisfactory, in consideration of the degree of uncertainty attending photometric operations.

(b)—Second order lenticular apparatus for a fixed light, with a catadi-optric accessory part.*

Lustre or brilliancy of the dioptric drum $=$ ·········· 160 burners.

Upper part, or cupola $=$ ···················· 76

Lustre or brilliancy of the catadioptric zones—lower

    zones $=$ ································· 28

Total lustre or brilliancy in burners················ 264

Calculation of useful effect.

264 burners $\times$ 20,000$'=$ ······················ 5,280,000

From which must be deducted for the space of one-sixth

  of the circumference, which is not supplied with the

  catadioptric zones below the lenticular drum········ 93,333

Remainder of the useful effect $=$ ·················· 5,186,667

Economical effect $= \dfrac{5,186,667}{500}$ grammes $= 10,373$.

If the apparatus illuminate only five-sixths of the circumference of the horizon, a reflector is placed on the side of the earth; the useful effect then becomes reduced to 4,706,667, and the economical effect to 9,413.

For an illumination embracing only three-fourths of the circumference of the horizon, the useful effect is then reduced to 4,120,000, and the economical effect to 8,240.

(b¹)—Second order revolving lenticular apparatus, with catadioptric accessory part.

The lustres or brilliancies measured upon half of the amplitude of the luminous beam emitted from a lens of 80 centimetres in height, occupying a space of 30 degrees.

⚬ For the second order (and with greater reason for the inferior orders) we dispense with the use of the *concave mirrors* for the subsidiary part of the apparatus. The increased useful effect of that part which is *always fixed* is, above all, important in the revolving dioptric drum apparatus to prevent, at a certain distance, the complete disappearance of the light or positive eclipses.

Azimuths.	Brilliancy or Lustre.		Angular distances.	Products.	Observations.
	Answering to the divisions.	Means.			
	DURNERS.				
210′	3 (?)	32 (?)	30′	960	
180′	61	167. 5	30′	5,025	
150′	274	383	30′	11,490	
120′	492	699. 5	30′	20,985	
90′	907	920	30′	27,600	
60′	933	1,006. 5	60′	60,390	
0′	1,080				
				126,450	

Useful effect of a lens $2 \times 126,450 = 252,900$.

Useful effect of 12 lenses of the drum $\cdots\cdots\cdots\cdots\cdots\cdots$ 3,034,800

Useful effect for catadioptric zones $\cdots\cdots\cdots\cdots\cdots\cdots\cdots$ 1,986,667

Total useful effect $= \cdots\cdots\cdots\cdots\cdots\cdots\cdots\cdots\cdots$ 5,021,467

Economical effect $= \dfrac{5,021,467}{500}$ grammes $= 10,043$.

(c)—Lenticular apparatus for a fixed light of the third order (larger size) with catadioptric accessory part.

Brilliancy of the dioptric drum $= 50$ burners; brilliancy of the catadioptric dome or cupola $= 20$ burners; total $= 70$ burners.

Useful effect $= 70$ burners $\times 20,000' = 1,400,000$.

Economical effect $= \dfrac{1,400,000}{190}$ grammes $= 7,368$.

If the apparatus illuminate only four-fifths of the circum-

ference of the horizon, the useful effect is reduced to $\cdots$ 1,120,000

To which add for spherical reflector $\cdots\cdots\cdots\cdots\cdots\cdots$ 40,000

Total useful effect $= \cdots\cdots\cdots\cdots\cdots\cdots\cdots\cdots$ 1,160,000

Economical effect $= \dfrac{1,160,000}{190}$ grammes $= 6,105$.

Catadioptric apparatus of the third order (smaller size) for a fixed light.

Brilliancy of the apparatus $= 31$ burners.

Useful effect $= 31$ burners $\times 20,400' = 632,400$.

Economical effect $= \dfrac{632,400}{115}$ grammes $= 54,999$.

If the apparatus illuminate only three-fourths of the circumference of the horizon, the *useful effect* will be reduced to 474,300, and the economical effect to 124.

It must bo remembered that this apparatus may be illuminated with a decided advantage by means of an *ordinary Argand lamp with one wick*, with which we will obtain the following results:

Brilliancy of the apparatus = 25 burners.*

Useful effect = $25 \times 20,400' = 510,000$.

Economical effect = $\dfrac{510,000}{60}$ grammes = 8,500.

Fourth order catadioptric apparatus for a fixed light (larger model.)

Brilliancy of the apparatus = 15 burners.

Useful effect = $15 \times 20,400' = 306,000$.

Economical effect = $\dfrac{306,000}{60}$ grammes = 5,100.

If the apparatus illuminate only three-fourths of the horizon, the useful effect will become reduced to 229,500, and the economical effect to 3,825.

Fourth order catadioptric apparatus (smaller model) for a fixed light.

Brilliancy of the apparatus = 9 burners.

Useful effect = $9 \times 20,400' = 183,600$.

Economical effect = $\dfrac{183,600}{45}$ grammes = 4,080.

If the apparatus illuminate only three-fourths of the circumference of the horizon, the *useful* effect becomes reduced to 137,700, and the economical effect to 3,060.

General observations upon photometric measures.

The photometric results from which I have deduced the values of the *useful effects* and of the *economical effects* of the various orders of *catoptric* and *dioptric* apparatus should only be considered as simple approximations. New experiments upon apparatus of the same kind

---

° The catadioptric apparatus of 50c. (19.69 inches) diameter, illuminated thus by an ordinary constant level lamp, burning 60 grammes (2 ounces 1.9 drachm avoirdupois) of oil per hour, offers a very advantageous combination for the illumination of the entrances to ports and roadsteads, with ranges of 12 to 15 nautical miles. *One keeper is sufficient to perform with ease all the service; and the ordinary annual expense does not exceed (in France) one thousand to eleven hundred francs* ($187.50 to $206.25.)

would show, without doubt, very remarkable differences, not only because of the greater or lesser precision which the construction of the optical pieces may present, and of the difficulty of obtaining the identity of brilliancy, in the *unity of light*, but also in consequence of the nature of the photometric experiments, where it becomes necessary to estimate by *the eye* the equal intensities of shadows produced by light, often of different tints.

These approximations are always sufficient to establish results exact enough between the effects of the different illuminating apparatus to be compared, and give much more correct ideas as to their relative values than can be deduced from observations upon the *absolute range* of the lights.

Service of the Catoptric lights.

The *fixed catoptric lights* established upon *land*, and illuminated by a small number of reflectors, say from six to eight, may be attended by a single keeper, especially if he be assisted by his family. If the apparatus be a revolving one, or if the number of reflectors be greater than six or eight, it will become, ordinarily, necessary to employ two keepers, as well to superintend the light during the night, as to maintain in a proper manner the illuminating apparatus and fixtures. If the light be placed in an isolated position at sea, three keepers at the least are necessary to assure regular attendance. In such cases, these agents (keepers) are relieved at regular intervals, and permitted to return to the continent for fifteen days or a month.

Service of the Lenticular lights.

The lenticular lights of the fourth order, and of the third order smaller model, require but a single keeper to superintend them, except in cases when situated at isolated points at sea.

Two keepers are required to superintend the lights of the third order larger model, and the second order, in consequence of the use of the mechanical lamps in those lights.

Three keepers are required for the superintendence of lights of the first order.

For lights of the second and third orders, in isolated positions at sea, it is necessary to employ *three keepers*, and *four keepers* for lights of the first order similarly situated.

COMPARISON OF THE TWO SYSTEMS OF LIGHTS.

Parallel between the Catoptric and Dioptric lights.

The preceding developments have appeared to me indispensable, as preliminaries to the establishment of a parallel between the two systems of illuminating light-houses.

I will consider the two systems under the following heads:

1st. The *absolute useful and economical effects.*

2d. The *first cost of the establishment*, and of *the repairs* and *maintenance.*

3d. The *facility* and *safety* of the service.

Apparatus of the fourth order, smaller model.

SEC. 1. *Absolute useful and economical effect of the illuminating apparatus.*—The *brilliancy* of a *catadioptric apparatus* of 30c. (11.8 inches) interior diameter, illuminated by a lamp burning 45 grammes (*one ounce, nine and four-tenths drachms avoirdupois*) of oil per hour, has been found equal to eight or nine burners.

The *brilliancy* of a "*sideral*" *reflector*, illuminated by a lamp burning 50 grammes (*one ounce, twelve and twenty-five hundredths drachms avoirdupois*) of oil per hour, is equal to four burners. This brilliancy of the first is at least *double* that of the second.

The *useful effect* of the *catadioptric apparatus* of the fourth order, illuminating three-fourths of the circumference of the horizon, is represented by 137,700. The *useful effect* of the "*sideral*" *reflector* is equivalent to 68,400. Comparison of the second to the first = 1 to 2. Economical effect of the catadioptric apparatus = 3060. Economical effect of the "sideral" reflector = 1368. Comparison of the second to the first = 1 to 2.24. If we take for a term of comparison the ordinary reflector, (*à coquille plate*,) the superiority of the catadioptric apparatus will be still more decided. With regard to the *concave parabolic reflectors*, or "*photophores*," I will not introduce them into this parallel, in consideration that they can only serve in isolated cases for the ordinary illumination of the entrances to harbors, upon an amplitude of not more than twenty degrees.

Apparatus of the third order, smaller model.

The apparatus of the third order, smaller size, illuminated by an ordinary *fountain, or constant level lamp*, carrying one burner and one wick, consuming 60 *grammes (two ounces one and nine-tenths drachm avoirdupois)* of oil per hour. or one burner with *two wicks,* consuming 115 *grammes (four ounces and one drachm)* of oil per hour, have no equivalents in the *catoptric apparatus* in use at present.

A "sideral" apparatus, of the same useful effect, would be of a dimension which would render the construction of it very difficult and very expensive, and would require, relatively, a very great consumption of oil. Neither could we supply it in a proper manner by the use of parabolic reflectors, except they were made expressly for the purpose, of very small dimensions, to allow a sufficient number to distribute properly the light upon three-fourths of the horizon. It is evident, moreover, that that embarrassing combination would require a consumption of more than 200 grammes (7 ounces) of oil per hour.

The old light of Cette. (Herault.) provisionally illuminated by a catadioptric apparatus of 50 centimetres (19.69 inches) diameter. with a lamp of two concentric wicks, is easily seen, in ordinary weather. at the distance of 15 to 18 nautical miles. although it is given *a range* in the official table of only 12 nautical miles.

Apparatus of the third order. larger model.

The brilliancy of catadioptric apparatus of the third order. one metre (39.38 inches) in diameter in the interior. illuminated by a mechanical lamp of double wick, burning 190 grammes (6 oz. 11.35 drachms) of oil per hour, has been found = 70 burners. I will suppose. moreover. that it embraces only four-fifths of the horizon.

To illuminate. by means of reflectors. the same angular space of 288°. with an effect of light *about equal.* it will be necessary to employ 14 parabolic reflectors (photophores) of 57.5c. (10.8 inches) diameter. each burning 35 grammes (1 oz. 3.8 drachms) of oil per hour. The brilliancy in the axis of each of the 14 reflectors will be about = 66 burners. The brilliancy in the intervals. the least illuminated, will be = 36 burners. The useful effect will be represented by $14 \times 62.160 = 870.240$. But the useful effect of the catadioptric apparatus has been found to be = 1,160,000. In this case. notwithstanding the very great difference in the consumption of oil, the dioptric or lenticular apparatus is superior, in useful effect, to the catoptric apparatus. The economical effect of the first apparatus is represented by 1786 : comparison = 1 to 3.44 ; that is to say, without estimating the expenditure of oil by *unity of light,* the lenticular apparatus will be nearly three and a half times more advantageous than the catoptric apparatus. With regard to the effective expenditure of oil, it will be in the proportion of 190 grammes to 14×35 grammes, or of 1 to 2.6.

Apparatus for a fixed light, second order.

The brilliancy of a catadioptric apparatus of the second order, having an interior diameter of 1m. 40c. (4 feet 7.13 inches) illuminated by a mechanical lamp of three concentric wicks, burning five hundred grammes (17 oz. 10.5 drachms) of oil per hour, has been found to be $= 264$ burners. Let us suppose it only embraces three-fourths of the horizon. To obtain an effect about equal in an angular space of $270°$, it will be necessary to employ at least thirty-four parabolic reflectors, having fifty centimetres (19.69 inches) diameter. The brilliancy in the axis of each of these thirty-four reflectors will be $= 270$ burners. But the lustres in the intervals are only $= 148$ burners. The useful effect of the catadioptric apparatus will be 4,120,000, and that of the thirty-four reflectors by $34 \times 103,680 = 3,525,120$. The comparison between the absolute expenditure of oil will be $= \dfrac{34 \times 42}{500} = 2.86$ to 1; and the comparison of the quantity of oil expended by unity of light $= \frac{34 \cdot 42}{6240} = 1$ to 3.33; thus, in this last respect, the lenticular apparatus will be three and one-third times as advantageous as the catoptric apparatus.

Revolving apparatus of the second order.

The *maximum brilliancy* or lustre of the revolving apparatus of the second order, with twelve lenses, has been found to be $= 1,184$ burners, as follows:

1st. Brilliancy in the axis of a lens $= \cdots\cdots\cdots\cdots$ 1,080 burners.
2d. Brilliancy in the fixed catadioptric zones $= \cdots$ 104   "

Total amount of brilliancy $= \cdots\cdots\cdots\cdots$ 1,184   "

And the *minimum* corresponding to the eclipses, is equivalent to one hundred and four burners.

To construct a catoptric apparatus, possessing an equivalent effect, without multiplying beyond bounds the reflectors, (photophores,) it will be requisite, without doubt, to take those with diameters of fifty-five centimetres to sixty centimetres (21.66 inches to 23.63 inches;) but for want of sufficiently precise data as to their lustres,

I will suppose the employment of reflectors of fifty centimetres (19.69 inches) diameter, which give, in their axis, a lustre equal to two hundred and seventy burners. I will suppose, also, that the frame has six faces, each fitted with four of these reflectors. I will admit, lastly, as the succession of flashes ought to be the same in both systems, that the lenticular apparatus makes its revolution in six minutes, and the catoptric apparatus in three minutes. The maximum lustre of the catoptric system will be equal to $4 \times 270 = 1,080$ burners. The amplitude of the lustres of the reflectors being, moreover, of sixteen degrees at most, there will be six angles of forty-four degrees each almost entirely obscured; and the length of these eclipses will be twenty-two seconds, while the length of the flashes will be only eight seconds.

Without pressing further the comparison of the effects of the two kinds of apparatus, we will perceive, without doubt, the evident advantages of the dioptric or lenticular combination, which, in fine weather, will not present an *absolute eclipse* at a less distance than fifteen to eighteen nautical miles. If we now consider the expenditure of oil, we will find—

First. That they are as $24 \times 42 : 500$, or :: 1 to 2 ;

Second. That the economical effects will be as $2,460 : 10,043$, or as 1 to 4.07.

Thus the lenticular apparatus will be *four times* as advantageous as the catoptric or reflector apparatus.

I admit that this result might be modified by the employment of larger reflectors ; but as their divergence would be less, the length of the flashes would be diminished, and consequently that of the eclipses would be augmented too much.

Let us remark before proceeding further, that in employing 25 parabolic reflectors (photophores) of 50 centimetres (19.69 inches) diameter for such an apparatus, we reach the utmost possible limit, without admitting the employment of lanterns of a size beyond all proper bounds ; and we may also affirm that very few of the catoptric lights, considered as lights of the *first order*, equal the lenticular lights of the same character of the *second order*.

Apparatus of the first order.

With the view not to multiply unnecessarily the comparisons, I will omit the details relative to the first order of the catadioptric combination, which is the most advantageous; and I will suppose, in consequence, the lenticular apparatus fitted in its accessory part with eleven zones of fixed mirrors. Now, we have found that the total lustre or brilliancy of an apparatus of that kind is equal in all its azimuths to 480 burners. But it will be practically impossible to obtain a like effect in the catoptric system, without having recourse to the employment of 36 reflectors with apertures (diameters) of 60 centimetres (23.63 inches.)

The difficulty becomes still greater if it be necessary to attain with these parabolic reflectors the effect of a revolving lenticular light with eight *large lenses*, the lustres or flashes of which exceed 4,000 burners of the Carcel lamp.

Let us limit ourselves, then, without entering into more full details, to the observation that the *economical effect* of a fixed light of the first order, illuminating three-fourths of the horizon, is to the *economical effect* of a light composed of parabolic reflectors ("photophores") of 50 centimetres (19.69 inches) diameter as 10.080 to 2.469, or as 4.08 to 1. That is to say, that the first will be (as to the expense of oil only) four times as advantageous as the second.

Apparatus for lights varied by flashes.

I will only here refer to the apparatus for *lights varied by flashes*, (otherwise called of *short eclipses*,) for the purpose of remarking that the catoptric or reflector system is not susceptible of producing, without great difficulty, that combination which unites to the permanence of fixed lights the advantage of presenting a very decided character.

SEC. 2. *Expenses of first establishing and of maintaining light-houses.*— So far, I have compared the *economical effects* of the different kinds of apparatus only with reference to the expenditures of oil; but it will not fail to be objected that the advantages which I point out in favor of the lenticular system ought to be very greatly diminished in a fiscal point of view, in consequence of the high price of the apparatus, and the excess of keepers necessary to their service. To appreciate the value of that objection, I will compare several lights of both systems, taking into the account all the expenses to which they could be subjected, and for the first example I will take the harbor lights.

HARBOR LIGHTS.

A harbor light, placed at the entrance to a port, being often exposed to the force of the sea, ought (to insure proper attendance, &c.)

to be erected securely in a bronze lantern, fitted upon the summit of a small tower constructed of masonry, and sufficiently spacious for the purposes of illumination and attendance at all times, and in every description of weather.

That being established, we may state the expenses for the first establishing for both systems, (in France,) as follows :

1. *Small "sideral" light.*

Small tower in masonry····························	5,000 fr.
Octagonal lantern, 1m. 40c. (4 feet 7.13 inches) in diameter..	2,650
"Sideral" reflector, with subsidiary pieces·············	500
= $1,528 12.	8,150
Interest at 5 per cent·····························	407 fr.
Annual expense of the service·····················	800
= $226 31.	1,207

2. *Catadioptric harbor light, (smaller model.)*

Tower and lantern······························	7,650 fr.
Catadioptric apparatus ·························	1,531
= $1,721 43.	9,181
Interest at 5 per cent·························	459 fr.
Annual expense of the service·················	800
= $236 06.	1,259

The *useful effect* of the "sideral" (reflector) light having been found equal to 68,400—

Its *economical effect* will be represented here by $\dfrac{68,400}{1,207} = 57.$

The *useful effect* of the catadioptric light, illuminating three-fourths of the horizon, has been found to be equal to 137,700—

After the same manner, its *economical effect* will be $\dfrac{137,700}{1,259} = 109.$

The comparison of the *economical effects* of these two will be, then, in the proportion of 57 to 109, or 1 to 1.91.

Then, besides the advantage of a *double lustre* or *brilliancy*, the catadioptric apparatus, in an economical point of view, is nearly twice as advantageous as the catoptric apparatus.

Lights of the third order.

It is difficult to establish a precise comparison between the fixed lights of the third order in the old and in the new system, because we cannot obtain with the ordinary parabolic reflectors a passably equal distribution of light, without multiplying these reflectors to such a number as would require a much greater expenditure of oil than could be allowed for lights of that class.

I will merely observe that I have every reason to believe, from the indications contained in the table of light-houses in the United States, that among all the lights of that country illuminated by reflectors, the diameters of which do not exceed sixteen English inches, there are very few whose *useful effect* is superior, or even equal to that of a catadioptric light of the third order, larger model.

[Revolving lights of the second order]

1st. Catoptric light.

In the comparative examination of the *useful and economical effects* of the two systems of maritime illumination, I have supposed the revolving catoptric apparatus of the second order, composed of twenty-four parabolic reflectors of fifty centimetres (19.69 inches) of aperture distributed equally upon the six faces of prismatic frame. I will suppose, also, that the lantern is three metres (9 feet 10.14 inches) in diameter, the same as required for a lenticular apparatus of the second order, but with a little less height. I will admit, moreover, the same perfection in the works—that is to say, that the frame and sashes, constructed of iron, are covered with bronze exteriorily; and that the roof or dome of the lantern is of copper, the glazing of glass of eight or ten milimetres (0.39 to 0.31 of an inch) in thickness, &c.

With regard to the buildings. the only remarkable difference will consist in, that the catoptric light-house will require room to lodge two keepers, while I will estimate for three in the lenticular light-house. That being stated, the necessary expenses to be incurred in the first system may be approximately estimated as follows :

Light-house tower, &c····································	50,000 fr.
Octagonal lantern, three metres (9 ft. 10.14 inches) in diameter, dressed in bronze·······················	10,500
Illuminating apparatus, consisting of twenty-four reflectors, fifty centimetres (19.69 inches) in diameter, revolving frame, rotary machinery, &c···············	12,500
Total first cost··························	73,000
=$13,687 50.	
Interest at five per cent·························	3,650 fr.
Annual expense of the service, estimating the oil at the highest price ····································	5,000
Total annual expense······················	8,650
=$1,621 87.	

2d. *Revolving lenticular apparatus of the second order.*

Light-house tower, &c····························	60,000 fr.
Lantern and subsidiary fixtures·····················	12,500
Catadioptric lenticular apparatus and mechanical lamps·	33,000
Total first cost ·························	105,500
=$19,781 25.	
Interest at five per cent························	5,275 fr.
Annual expense of the service ···················	5,800
Total annual expense ····················	11,075
=$2,076 56.	

The useful effect of twenty-four parabolic reflectors of fifty centimetres (19.69 inches) diameter will be equal to $24 \times 103,680 = 2,488,320$.

The economical effect will be represented by $\dfrac{2,488,320}{8,650} = 288$.

The useful effect of a lenticular light of the second order, with twelve revolving lenses, has been valued at 5,021,467.

The economical effect will then be represented by $\dfrac{5,021,467}{11,075} = 453$.

The *economical effects* of these two lights will then be in the proportion of 288 to 453, or of 1 to 1.6. From whence it results definitely, that the lenticular light of the second order will be *more than one and a half times* as advantageous as the catoptric light, which we may without doubt consider as being of the first order, and the *useful effect* of which, nevertheless, could not be equal to but *the half* of the *useful effect* of the first.

Lights of the first order.

I find myself arrested in the attempt to make a comparison of the catoptric and dioptric lights of the first order, for the reason that we could not, without increasing beyond all proper limits the number of reflectors and the dimensions of the lantern, construct a catoptric apparatus to produce a sufficiently powerful effect to be assimilated to a dioptric apparatus of the first order. I mention this in this place merely that it may not be forgotten.

From the preceding details we may conclude :

1st. That the lights fitted with the dioptric apparatus present a variety in their power and in their effects, and may be made to produce an intensity of lustre which renders them of an interest in a nautical point of view, incontestably superior to those fitted with the catoptric apparatus.

2d. That if we take into account the first cost of construction, and the expense of their maintenance, we will find, in respect to the effect produced, the new system, (dioptric) is still from *once and a half to twice as advantageous as the old*.

FACILITY AND SECURITY OF THE SERVICE OF THE LIGHT-HOUSES OF THE OLD AND NEW SYSTEMS.

After having balanced the advantages relative to the two systems of lights in view of their *useful* and *economical effects*, I ought to consider them with reference to their *security* and the *facility* with which they are served.

I will reproduce, upon this subject, the observations inserted in a memoir of the 20th of April, 1830, in which I replied to the questions which were addressed to me by the government of Sweden and Norway, in relation to the necessary measures to be taken to improve the lighting of its maritime coast.

"The service of lenticular lights is, in the aggregate, less laborious than that of the reflector lights. The first, demand at all times during the night the unremitted attention of the keeper. If, for example, the central lamp should become extinct during the absence of the keeper from the lantern, or while he is asleep, the horizon of the light would remain some hours plunged in total darkness, and the greatest objection which has been urged against our new system of illumination is the fear of such accidents. Happily, an experience of *seven years* has dissipated that fear, and the lenticular lights have been distinguished up to this time by the regularity of their service. However, every precaution has, besides, been taken to replace promptly the lamp or its burner in case of extinction. The extreme simplicity of the day duty compensates the keepers for that to which they are subjected during the night. To snuff and replace the wicks, renew the oil, sweep the chambers of the lantern and the stairs of the tower, dust the apparatus, and sometimes wash with a little spirits of wine the tarnished spots upon it, and, lastly, to wipe dry the glass of the lantern—such is the principal daily duty which is divided between the keepers of the new lights, and which rarely occupies them more than two hours."

Opinions thus expressed *fifteen years since*, based upon an experience of seven years, have been greatly strengthened up to the present time, embracing a period of *twenty-two years* since the establishment of the Cordouan light, and sustained by the results daily offered of more than *one hundred* lenticular lights of the three first orders, established along the coasts as well of France as of different foreign powers.

In this important point of view, then, the question seems to be irrevocably settled, and I will only add a few considerations relative to the application, more or less extended, which may be made of the new system of illumination, to the vast maritime coasts of the United States.

1st. It has been objected that it would require too great sacrifices to be made to procure in that country keepers possessing the amount of intelligence requisite for the superintendence of lenticular lights.

2d. That from distant points or stations the necessary repairing and renewing of the mechanical lamps would be attended with great difficulty.

I will reply, with regard to the *keepers*, that the difficulty of obtaining proper persons to fill these subaltern stations appears to be most singularly exaggerated. In France they belong almost always to the class of *ordinary mechanics*, or laborers, who make from 1.50 francs to 2.50 francs (27 to 46 cents) per day.

Eight or ten days will suffice, ordinarily, to instruct a light-keeper in the most essential parts of his duty, receiving lessons from an instructor conversant with all the details of the service ; and two instructing officers will be sufficient to prepare keepers for all the lenticular lights which could be successively established upon the coasts of North America. The information thus imparted, would never be lost ; and these officers might, besides, be aided by foremen or assistants, who could supply the place in case of necessity. In defence of this assertion, I will cite the example of the administration of Norway and Sweden, which, after having obtained the assistance of a French agent to put up the apparatus of the two first lenticular lights, which were sent from Paris in 1832 and 1836, has provided since, without any foreign assistance, for the placing, as well as the organization of the service, of all the lights of the new system which it has successively established.

With reference to the eventual repairs of the mechanical lamps, it is to be considered—

1st. That, in consequence of the great strength of the pieces of which the new model of mechanical lamps are composed, they will perform well for a number of years without requiring anything more than a proper attention to their cleanliness.

2d. That the ordinary assortment of a dioptric light-house comprises three of these lamps, which afford a sufficient guarantee against the chances of accident ; and besides, we may, by increasing a little the mean expense, increase the number to *four* under some circumstances, as an exception to the general rule.

3d. That the repairs of the implements under discussion may be easily made by all the clock or watchmakers, or other mechanicians, to whom we have recourse for repairing the revolving machinery of light-houses.

I conclude with the remark, that if it be determined to multiply the application of the new system of maritime illumination in the United States, it seems to me that it will be expedient to *engage one of the foremen employed in the manufactory of our mechanical lamps to go to, and remain in the country for several years.* By that measure, which would be attended with very little expense, all the difficulties which might present themselves at first in establishing lenticular lights would be removed, and the perfect regularity of the service of these new establishments would be insured.

· · LEONÓR FRESNEL.

PARIS, *December* 31, 1845.

FRANCE.

Extract from Parliamentary report, 1834.—*Light-house service of France.*

The light-houses, for the most part, existing on the coast of France before the revolution, had been built and were managed by commercial bodies.

A law of the National Assembly, dated the 15th September, 1792. centralized the light-house, beacon, buoy, and sea-mark service, by placing it under the superintendence of the Minister of Marine, and by charging the Minister of the Interior with the execution of the works agreed upon for this service by the two departments.

A consular order, dated the 11th June, 1802, confirmed the law of 15th September, 1792, concerning light-houses.

An imperial decree of the 7th March, 1805, caused the light-house, buoy, sea-mark, and beacon service to be attached to the official duties of the Minister of the Interior, and from that time they were placed under the immediate direction of the commissioners of roads and bridges. This decree, nevertheless, requires the Ministers of the Interior and of Marine to advise with each other in reference to the establishment of any new light-houses or sea-marks; and this arrangement gave rise to the light-house commission, which was established in 1811.

In 1825, at which period the system of lenticular lights was definitely adopted for lighting the coast of France, this commission is found to have been composed as follows:

M. Becquey, councillor of State, director general of roads and bridges, president of the commission.

M. De Prony, inspector general of roads and bridges, member of the Institute.

M. Tarbé de Vaux Clairs, inspector general of roads and bridges, councillor of State.

M. Sparkin, inspector general of hydraulic works at the seaports.

M. Rolland, inspector general of naval architecture.

M. Halgar, rear admiral and councillor of State.

M. De Rossel, rear admiral, director of the repository of charts and plans of the royal navy, member of the Institute.

M. Beautemps Beaupré, Hydrographer-in-chief of the royal navy, member of the Institute.

M. Arago, Astronomer, member of the Institute.

M. Mathieu, do. do. do.

M. Fresnel, (late Augustin,) principal engineer of roads and bridges, member of the Institute, secretary of the light-house commission.

It was by this commission that all the projects and measures adopted since its institution for the improvement of the light-house and sea-mark service were examined and discussed, and one of its most important labors was the study and examination of the general system adopted in 1825, upon the report of Rear Admiral de Rossel, for lighting the coasts of France.

Present organization of the light-house service.

The organization of the light-house service of France, now in force, is briefly as follows:

This service is attached to the official duties of the Minister Secretary of State for the Department of the Interior, and is altogether under the direction of the Councillor of State charged with the general administration of the roads and bridges.

In each naval district, the prefect, principal engineer, assistant engineers, and the superintendents of ponts et chaussées, direct or supervise, in the sphere of their respective offices, all that relates to the management of light-houses, sea-marks, and beacons in the neighborhood.

The light-house service, considered collectively, embraces lighting, reparatory works, and the formation of new establishments.

Reparatory works.—The repair or restoration of light-houses, after being authorized by the director general, is, with the roads and bridges works, executed under the superintendence of the district administration.

New establishments.—In the formation of new establishments, the following routine is observed :

The engineers of the district, where the new edifice is about to be erected, make a draught of the plan, in conformity with the basis previously determined on by the light-house commission.

This plan is forthwith submitted to the commission, which confines

12

itself to the inquiry as to whether the wants of the service, nautically or otherwise reported on, and constituting the main objects, have been complied with.

It is then presented to the council of roads and bridges to receive their estimate, founded on the reports made of the architectural arrangements of the system of building, and of the calculated expense. After receiving the approbation of the director general of roads and bridges, and of the Minister of the Interior, the plan is sent to the prefect of the district, who proceeds to the public adjudication (contract) of the works, and intrusts the engineers with the execution of them.

The lamps and light apparatus are made at Paris, under the care of the engineer-in chief, secretary to the light-house commission.

The establishment of new light-houses is announced to the public two or three months beforehand, by means of bills and advertisements inserted in the maritime newspapers.

The administration, moreover, publishes annually a summary description of the light-houses and lights on our coasts, and causes five or six thousand copies to be distributed among French and foreign navigators.

The whole of the expenses (with the exception of the cost of lighting a very small number of lights of purely local interest) connected with the light-house service are supplied from the public treasury. The administration of the customs of France does not levy any special light-house duty upon maritime commerce. This duty, which was abolished by the law of the 18th October, 1793, is at the present time compounded with the tonnage duty which all vessels pay upon their arrival in port.　　　　LOR. FREINEL,

Secretary of the Light-house Commission.

PARIS, *April* 30, 1834.

List of the French Light-house Commission in 1851, instituted April,
1811.

The Minister of Public Works, or in his absence the under Secretary of State presides during the sittings of the commission.

M. Arago, commander in the Legion of Honor, representative of the people, member of the Institute and of the bureau of longitudes·

M. Mathieu, Knight in the Legion of Honor, member of the Institute and of the bureau of longitudes.

M. Mathieu, Commander in the Legion of Honor, rear-admiral.

M. De Hell, High officer in the Legion of Honor, rear-admiral, *en retraite.*

M. Leroux, Commander in the Legion of Honor, general inspector of maritime engineering.

M. Tretté de Laroche, Officer in the Legion of Honor, divisionary inspector of bridges and roads, charged with the general inspection of maritime works.

M. Fresnel, Officer in the Legion of Honor, divisionary inspector of bridges and roads, *en retraite.*

M. Reynaud, Knight in the Legion of Honor, chief engineer of bridges and roads, secretary.

List of the Trinity-House Board, London.—Elder Brethren in 1851.

Duke of Wellington, K. G., Master.
Captain Sir J. H. Pelly, Deputy master.
Captain Aaron Chapman.
Right Hon. Lord Viscount Melville, K. T.
Captain Robert Welbank.
Captain John Hayman.
Captain Henry Nelson.
Admiral Sir T. Byam Martin, G. C. B.
Captain Charles Weller.
Right Honorable Sir J. R. G. Graham, Bart.
Right Honorable Earl of Minto, G. C. B.
Captain Frederick Maden.
Admiral Sir Charles Adam, K. C. B.
Captain Stephenson Ellerby.
H. R. H. Prince Albert, K. G.
Captain George Probyn.
Captain William Pixley.
Captain Charles Farquharson.
Captain Robert Gordon, R. N.
Captain William E. Farrer.
Captain Henry Bonham Box.
Right Honorable Earl of Haddington.
Most Honorable Marquis Dalhousie.
Captain John Shepherd.

Captain Edward Foord.
Captain Gabriel J. Redman.
Right Honorable Lord J. Russell, M. P.
Captain John Fulford Owen.
Capt. David James Ward.
Right Honorable H. Labouchere, M. P.
Capt. Wm. Pigott.
Jacob Herbert, Esq., Secretary.

TRANSLATION.

[Ministry of Public Works—Second Bureau—Navigation—Light-houses.]

Control of the consumption of oil, and of the condition of the different supplies.

PARIS, *March* 17. 1845.

SIR: The great expense incurred by the administration for the purpose of carrying the illumination of our maritime coasts to a degree of perfection which will insure the safety of navigators, will not fulfil perfectly that object if the service is not, on the part of the engineers, an object of their special care and superintendence; yet, however active they may be, it will be conceded that they may fail very often in the accomplishment of their object. We may, doubtless, by a minute inspection of the apparatus, the implements, and the supplies, judge if the light-keepers have at their command all the elements requisite to an efficient service, and if they are employed with all proper care. We may equally discover, either from observations made at a distance, or unexpected visits, if the perturbations observed in the illumination of the light-houses arise from the negligence of the keepers, or from their want of capacity; but these means only afford us an idea of the mean state of the illumination, and from which we may accidentally discover some isolated facts; but that will not suffice for the proper direction of a service of public safety, the normal condition of which should never be permitted to be interrupted.

To obtain a more positive and a more efficient control, without recourse to an unremitted supervision, which would require a large number of special agents, it has appeared to me that the most direct and the surest means was to require the principal light-house keepers to keep a regular account of the oil consumed each night for the illumination of their lights. Doubtless there is reason to fear that these accounts will not always be correct; but we shall, sooner or later, detect false declarations by the details of the monthly tables, of which we will speak hereafter.

This new plan, already put into practical operation with perfect success in some light-houses, is only applicable to the lenticular apparatus, illuminated by lamps with multiple wicks. These lamps require in their management unremitted attention and care, to insure a proper development of their flames. We have, also, generally to encounter the lazy tendency of the keepers in keeping habitually the flames of the lamps below their normal height.

This is a capital point, and is the hinge on which the whole light-house service turns.

The table to be filled by the principal keepers must be kept in conformity to the accompanying model, (A') divided into six columns, which comprise respectively the following indications:

First column, the day of the month.
Second column, the hour of lighting.
Third column, the hour of extinguishing.
Fourth column, the duration of the illumination.
Fifth column, the weight of the oil consumed.
Sixth column, the remarks.

The direct weight of the oil put each day into the reservoir of the lamp, and the residue which is drawn off after extinguishing the light, will present in the service a complication which may be avoided by means of a gauging table.

To construct that table we may proceed as follows: Refill the reservoir of the lamp by pouring one kilogramme of oil at a time into it; note at each time the number in millimetres corresponding to the depth of the liquid, and, in dividing by ten the successive differences, we will determine the numbers answering to the graduated weight of each hectogramme.

If the three or four lamps employed in the light-house have not their reservoirs of exactly the same shape, it will become necessary to construct a special gauging table for each lamp, which must, in that case, be distinctly marked, so that they may not produce mistakes. The reports of the consumption of oil for each month must be signed (at the end of each month) by the principal keepers, who will address them in triplicate to the deputy of the lighting contractor, or to the conductor of the arrondissement, according as the supplies are furnished by contract or the administration. One of the three copies will be retained by the deputy or conductor, and the two others, with the signature of that agent upon them, must be transmitted to the engineer of the arrondissement. Lastly, one of the two last must be sent to the engineer-in-chief, who, after having signed it with or without remarks, addresses it to the secretary to the commission of lights.

Independently of these tables, destined specially to control the nocturnal service of the lighting establishment, it is necessary to continue to require the principal keepers of the lights under the administration, as well as those under the contractors, to furnish monthly statements of the condition of those supplies the most essential to the illuminating service.

The accompanying model (B²) is used by the administration, and is applicable to the contracts, by substituting the approval of the deputy of the contractor for that of the conductor or agent of the administration. These latter reports must be made out in quadruplicate, disposed of in the same manner that the others were, and forwarded under the same cover.

This second precautionary measure will have the advantage of furnishing a base of verification for the daily consumption of oil, and of apprizing the engineers at the same time, more readily, if the light-houses are supplied at all times with all the objects necessary as well for the service of the illumination as the maintenance of the apparatus confided to the light-keepers.

To prevent these new measures from imposing any expenses upon the contractors not contained in the table of charges, the administration will furnish the lithographed sheets necessary for the keeping of the two different tables ; and they will be sent under cover to the engineers or conductors.

I pray you, sir, to have the kindness to communicate this circular to the engineers and conductors charged, under your orders, with the service of the light-houses, and require the strictest execution of all duties embraced in it.

Receive, sir, the assurance of my very distinguished consideration,

———— ————,

Under Secretary of State for the Public Works.

To Monsieur ————, *Engineer-in-chief at*

[Translated from the French.]

MEMOIR PRESENTED TO THE ACADEMY OF SCIENCES, JANUARY 8, 1844.

(Lětourneau & Cie., successors of Messrs. Soliel, sr., and François, jr., constructors of dioptric lights upon the system of Mr. A. Fresnel, ''Rue des Poissonnières, No. 24 près et hors la barrierè Poissonnières à Paris.'']

Note upon the catadioptric apparatus constructed by Mr. Francois, jr., for the Scotch light at Skerryvore.—(Commissaries, MM. Arago, Mathieu, Babinet.)

The lenticular apparatus imagined by M. Augustin Fresnel comprises, independently of the fixed or movable dioptric drum, an accessory part, destined to collect and direct towards the horizon the luminous rays which, issuing from the focal centre, pass above and below the lenses.

This accessory part has been in most cases formed by a system of fixed concave mirrors, arranged in horizontal zones, both above and below the lenticular drums.

In the two light-houses of Cordouan and Marseilles, the revolving dioptric drum is surmounted by a system equally movable, composed of eight lenticular panels arranged in the form of a truncated pyramid, and as many plane mirrors to convey towards the horizon the eight luminous beams of light emerging perpendicularly to the faces of the pyramid. A third combination, preferable to the two others in the double aspect of theory and practice, has been applied by the inventor to the small lenticular lights of twenty-five to thirty centimetres of interior diameter. In these apparatus, which, in consequence of their small size, are not adapted to the employment of mirrors, the accessory catoptric system has been replaced by a catadioptric system of rings or zones, in triangular section, producing total reflection.

The first apparatus of this description was constructed a short time before the death of M. Augustin Fresnel, by M. Tabouret, conductor of bridges and roads, attached to the special service of the light-houses.

The application of this system to the apparatus of the largest size ought, at that time, to have appeared almost impossible. Scarcely, in fact, could the shape of the dioptric rings of seventy-five to eighty centimetres in diameter for the large plano-convex lenses be obtained. As to the fixed dioptric drums, of diameters exceeding thirty centimetres, they were composed of cylindrical elements, the assemblage of which presented, in place of an annular, a polygonal system of sixteen sides for the apparatus of fifty centimetres diameter, (third order smaller model;) twenty sides for the apparatus of one metre in diameter, (third order larger model;) twenty-four sides for the apparatus of one metre forty centimetres in diameter, (second order;) and thirty-two sides for the apparatus of one metre eighty-four centimetres in diameter, (first order.)

A catadioptric polygonal system could, without doubt, be executed by the means employed for the polygonal dioptric system : but the adjustment of such a multitude of prisms of reflection, the positions of which could not be exactly regulated except when put up, presented an inadmissible complication.

It was necessary, for the practical solution of the problem, that the moulding and cutting of the large pieces of glass should be performed by an improved method. A manufacturer of mirrors at Newcastle-upon-Tyne, England, (Mr. Cookson,) placed, with regard to the making experiments upon the subject, in a singularly favorable position, because of the facilities of every description which his vast establishment afforded him, made the first attempt, in 1836, to construct the dioptric drums of the first order entire, which up to that time had been formed in prisms of thirty-two panels. The results of these first essays, without being fully satisfactory, stimulated the zeal of the French artists, who were devoting themselves to the fabrication of the lenticular apparatus ; and very soon after, we obtained dioptric drums of nearly two metres in diameter, executed entire, with a precision which increased the useful effect of that principal part of the apparatus about one-fourth.

From that time the project of constructing upon a large scale the catadioptric apparatus was permitted to recommence, with some chances of success. However, the considerations of expense and little prospect of profits, &c., were very discouraging.

Nevertheless, the able Scotch engineer charged with the construction of the Skerryvore light-house (Mr. Alan Stevenson) devoted himself with zeal and assiduity to the idea of crowning that monument (which cost about two millions of francs) with the most perfect illuminating apparatus which it was possible to construct in the present state of the arts and sciences.

The programme was adopted by the commission of Scotch light-houses, and it was determined that the Skerryvore rock should be marked by a catadioptric apparatus of the first order, the optical parts of which should be constructed at Paris.

A correspondence followed upon the subject between Mr. Alan Stevenson and Mr. Leonor Fresnel, the engineer secretary of the commission of French lights. The latter first calculated the elements of, and caused to be constructed, as a first attempt, two catadioptric apparatus of one metre diameter, (third order,) one of which, constructed at the establishment of Mr. Henry Lepaute, has illuminated for some months past the entrance to the port of Gravelines ; and the other, constructed by M. Francois, jr., is destined for the light-house which is being erected at the mouth of the Abervrach, upon the northwest coast of Finisterre.

In spite of the complete success of the first experiment, the fabrication of the reflecting rings of the first order always presented itself as a grave and perilous enterprise.

Even the secretary to the commission of French lights, in remitting to M. Francois, jr., the table of the centres and radii of curvature of the nineteen glass rings or zones which were required to form the catadioptric part of a first order light, deemed it his duty to insist that that artist should weigh well the engagement which he was about entering into with the administration of the Scotch lights. M. Francois, jr., did not hesitate a moment. He undertook resolutely a work of great public utility, which required in its accomplishment the overcoming of grave difficulties.

One may form some idea of these difficulties by a simple inspection of the table of radii of curvatures of the reflecting surfaces of the catadioptric rings, which vary from 6,816 metres to 8,749 metres.

The ring No. 1, which answers to the maximum radius, has two metres of exterior diameter. The two adjacent sides of the obtuse angle (of 117° 26' 42'') have respectively 92.380 millimetres, and 95.209 millimetres of length. The two refracting faces have been supposed rectilinear in the calculation ; but in consequence of the difficulty of executing with precision conic surfaces, we have (following the ingenious idea of the inventor) substituted for the two generating right lines two arcs of a circle of equal radius, (four metres,) taking care to turn them in an inverse sense, so that the convergence resulting from the convexity of one face was compensated by the divergence resulting from the concavity of the other face.

Each ring has been composed of four equal arcs.

These pieces were first run into a rough shape at the manufactory of Saint Gobain, in the moulds furnished by M. Francois, jr.

The first operation presented difficulties which would have discouraged one less determined, and one possessing a mind less fertile in resources.

Each rough ring was afterwards rubbed with grit or freestone, smoothed with emery, and polished with English red, upon a circle moved by steam machinery.

It may be conceived how many precautions are required in the perfect execution of an annular reflecting surface, which is cut or shaped by means of a rubber, grinding it with an oscillating arm or lever of eight metres seventy-five centimetres in length, and how much more care ought to be taken to study the means by which to insure the rigidity of this arm or lever, as well as the exactitude of the position and the fixedness of the centre of rotation.

Not only has this difficult problem been solved with perfect success, but it has been done without groping or wavering, without false movements, and without having to regret the loss of one ring broken upon the grinding circular frame.

After having been verified by the reflection of a red ball placed in their focus, the rings or zones were put together in panels.

To fulfil the requirements of Mr. Alan Stevenson, M. Francois, jr., divided his catadioptric dome or cupola into eight spindles, embracing, each, forty-five degrees. One of these spindles has been put under experiment twice at the observatory. Illuminated by a lamp of the first order, with four concentric wicks, burning from 670 to 700 grammes of oil per hour, this catadioptric panel presented a brilliant bar of light, which, after the mean of six observations of equal shadows, was equivalent to one hundred and forty burners of the Carcel lamp, burning forty-two grammes of oil her hour.

The catoptric cupola, which the new system replaces, is composed, ordinarily, of seven horizontal zones, containing each thirty-two concave mirrors. Its brilliancy appeared greater or lesser according as it was placed in the direction of the axis, or of the intervals of the mirrors, but the mean lustre corresponding to the useful effect has been found to be eighty-seven burners of Carcel.

Thus, then, the useful effect of the new crown is to that of the old one as 1.61 to 1.

It is to be presumed that the same results would be found to exist, or pretty nearly the same, for the part below the lenticular drum;

and as we have found forty-six burners for the mean brilliancy of the four lower zones of mirrors, we may count upon seventy-four burners for the brilliancy of the six corresponding catadioptric rings. The value of a fixed lenticular drum of the first order, with annular elements, being, besides equal to 360 burners, we may recapitulate by the following little table the approximation to which it refers :

Lustres or brilliancy, measured in burners of Carcel.

		First system.	Second system.
1. Fixed dioptric drum · · · · · · · · · · · · · ·		360 burners.	360 burners.
2. Accessory parts { Cupola · · · · · · · · · · · ·		87 "	140 "
{ Lower zones · · · · · ·		46 "	74 "
Total ·		493 "	574 "

Finally, the substitution of the prismatic rings in place of the mirrors of a first order fixed light, will augment the mean brilliancy eighty-one burners—that is to say, more than equal to the value of a light of the third class.

To that increase of the effect of sixteen and a half per cent. upon the total brilliancy, two capital advantages are joined : one, the equal distribution of light, and the other the stability of the reflecting power of the catadioptric rings.

Although the fiscal question is only one of a secondary consideration here, yet perhaps it may not be superfluous to say a word upon it in concluding.

The system of eleven zones of curved mirrors of a light
of the first order costs, including the subsidiary pieces, 6,000 francs.
The corresponding catadioptric system has been tendered for the price of · 20,000 "

Augmentation · 14,000 "

If, then, we take for example a light of the first order,
costing annually for illuminating, and the ordinary
service · 7,500 "
Add the interest on first cost of illuminating apparatus 1,500 "

9,000 "

we will find that the above calculated advantage of sixteen and a half per cent. would be equal to 1,485 francs, a larger sum than the interest upon the 14,000 francs, the excess of the price of acquisition.

Thus, then, considering the new system in a fiscal point of view only, we perceive that the augmentation of useful effect which it will produce will not be acquired at too high a price.

Mitchell's screw-piles and moorings.

SIRS : Permit us to present to your notice a brief description of our patent mooring, an instrument now well known in many of the principal ports and harbors of the United Kingdom, where they have been for a considerable time in constant use, and where we may add, they have had the unqualified approbation of every scientific and nautical person who has had occasion to consider the subject.

The mooring is constructed (as its name implies) on the principle of the screw, but differing essentially in form from that well known instrument; for while the spiral thread makes little more than one turn round its shaft, it is, at the same time, extended to a very broad flange, the hold which it takes of the ground being proportional with its breadth of disk.

Where it is necessary to provide against a very heavy strain, we have hitherto used moorings of three feet six inches diameter, and the principle is capable of still further extension.

A mooring of the above diameter presents a resisting surface equal to about ten square feet, whereas the palm of the largest anchor in the British navy does not exceed half that size ; and some estimate of its holding power may be formed, when it is shown that this broad surface can be screwed to a depth many times greater than that to which the palm of an anchor can ever descend.

The method of laying down this mooring is briefly thus :

A strong mooring chain being so attached to it as to allow the screw to turn freely without carrying the chain round with it, a powerful iron shaft is then fixed firmly on the upper part of the mooring, which is formed square for that purpose, fitting in the same manner as a key to a watch in winding it up : it is then lowered by the mooring chain, joint after joint being added to the shaft till the mooring has reached the ground. Eight levers of twelve feet in length are then applied to the shaft, in the manner of a capstan, when the operation of screwing the mooring into the ground commences.

Two boats or barges having been moored firmly, head and stern, close alongside each other, and the upright shaft rising between them about midships, the men place themselves at the bars or levers, and move round from one boat to the other, the two giving them a safe and convenient platform. By a simple contrivance, the levers are occasionally shifted upwards, as the screw and the shaft sink into the ground.

When the number of men employed can no longer force the screw round, the levers are removed and the shaft drawn out of the ground, leaving the mooring firmly embedded, with the chain attached to it; a buoy being shackled to the other end of the chain, the work is completed; the time required for the whole operation seldom exceeding a few hours.

These moorings have been placed in every description of ground—rock not excepted; and the qualities which entitle them to the patronage of the public are, perfect security to shipping, great economy, and an entire freedom from the many objections to which other moorings are liable. On these subjects, however, we shall not enlarge, but refer to the letters and testimonials hereto annexed, which have been selected from a great number of documents received on the subject.

Any communications addressed to Alexander Mitchell & Son, engineers, 2 Alfred street, Belfast, shall receive prompt attention.

We have the honor to be, sirs, your very obedient servants,

ALEXANDER MITCHELL & SON.

Lieuts. JENKINS and BACHE, *U. S. N.*

Extract from the report of Captain Canfield, Corps of Topographical Engineers, to Colonel Abert, dated October 1, 1851, *and printed as part of Senate document No.* 1, *of the session* 1851-'52.

As there is only one burner or lamp in this apparatus, it is evident that any accident by which the lamp was put out would leave them in total darkness; differing essentially in this respect from the common reflecting lights, where, if several lamps go out, there will still be a light of some kind. To make it certain that the lamp is kept burning, it is usual to keep a watchman constantly with the light.

As a substitute for the watchman, and that the keeper may know immediately of any accident of this kind, I have fixed a contrivance at Waugoshance, by which the the fog-bell will commence ringing as soon as the light is put out.

This is effected by making use of the expanding and contracting power of a copper tube, when heated and cooled. The tube is made to form a part of the chimney of the lamp.

The amount of the expansion under this heat is very small, (about one-twentieth of an inch.) But the expansive force being very great, I use a lever, with a short fulcrum running from the lamp at the centre to the side of the lantern, and increase the amount of the movement here ten times. To the end of this lever is attached a copper wire, which wire is also attached to another lever on the floor of the bell machine, forty feet below the lantern.

The movement by this second lever is again increased six times: so that the motion here amounts to full three inches.

This last lever, when the copper tube is heated by the lamp, is in a position to hold an iron hook, so that the hook will catch one of the spurs of the wheel which moves the clapper-shaft.

The catching of the spur, of course, stops the machine; when the light goes out, the copper tube of the chimney cools and contracts. The lever at the machine is raised, and the hook by its own weight swings clear of the spur of the wheel, and the machine moves on, and the bell rings: and unless it is stopped, continues to ring until the machine runs down.

This arrangement is found to answer perfectly, and is tested every morning when the light is put out.

EXTRACTS FROM TREATISES ON LIGHT-HOUSE ILLUMINATIONS, &c., &c., &c.

Extracts from Mr. Alan Stevenson's Treatise on Light-houses, &c.

[John Weale, London, 1850.]

Colza oil has been introduced in England into all the lights, whether catoptric or dioptric; but in Scotland its general use has as yet been confined to the dioptric lights, and such catoptric lights as revolve, and are not likely to be changed to the dioptric system. In the catoptric lights, the only reason for not making an equally extensive trial is the necessity for renewing all the burners, which require to be so constructed as to receive thick wicks of brown cotton; and it has, until lately, been considered prudent to proceed with some cauin changing the apparatus, so as to suit it for burning a patent oil, the circumstances attending the regular and extensive supply and the price of which, can hardly yet be fully known. The change is proceeding gradually from the use of spermaceti to that of the colza oil; and, in a few seasons, the whole will be completed, as nearly all the revolving catoptric lights have been altered; and the change on the fixed lights has been only delayed until they shall be converted to the dioptric system, so that one change may serve every purpose at once.   *   *   *   *   *   *   *

Great as Argand's improvement undoubtedly was, the value of the lamp alone as a means for the illumination of light-houses must be regarded as comparatively small. The primary object of a light-house is to give early notice to the mariner of his approach to the coast; and it is therefore necessary that the light be of such a kind that it may be seen at a great distance. Every one is practically acquainted with the fact that the rays proceed in all directions from a luminous body in straight lines; and if we could obtain a ball equally luminous in every part of its surface, it would give an equal share of light to every part of the inner surface of a hollow sphere, whose centre shall coincide with the centre of the ball. Again, if an opaque body were placed between the luminous ball and the hollow sphere, the part opposite that body would be deprived of the light by the interception of the rays, and no light would emerge from a hole bored in that part of the surface of the hollow sphere. The bearing of these facts is obvious; and no one can fail to perceive that in the case of a light-house illuminated by a single unassisted burner, a seaman could only receive the benefit of that small portion of light which emerges from the lamp in a line joining his eye and the centre of the flame. The other rays would be occupied partly, but in a very small proportion, in making the light visible in other parts of the horizon; while all the rest would be lost by escaping upwards into the sky, or downwards below the plane in which seamen can see a light-house. This state of matters would be little improved by increasing the number of burners, as the effective part of the light would only be augmented by the addition of an equally trifling portion of light from *each* burner. The small pencils of rays thus meeting at the eye of a distant observer, would form a very minute fraction of the whole quantity of light uselessly escaping above and below the horizon, and also at the back of each flame; and the wasteful expenditure of light would be enormous. By such a method, no practically efficient sea-light could ever have been obtained.   *   *   *   *   *

The best proportions for paraboloïdal mirrors depend on the objects which they are meant to attain. Those which are intended to give great divergence to the resultant beams, as in fixed lights, capable of illuminating the whole horizon at one time, should have a short focal distance; while those mirrors which are designed to produce a nearer approach to parallelism (as in the case of revolving lights, which illuminate but a few degrees of the horizon at any one instant of time) will have the opposite form. Those two objects may, no doubt, be attained with the same mirror, by increasing or diminishing the size of the burner; but that is by no means desirable, as any change in the size of a burner, which is found to be the best in other respects, must be considered as to some extent disadvantageous.

*　　*　　*　　*　　*　　*　　*

The large mirrors used in the Northern light-houses have about twelve-seventeenths of the whole light of the lamp incident on their surface; the rest escapes in the comparatively useless state of naturally radiating light. Several arrangements have been proposed for economising this light, which will be afterwards noticed.

The reflectors used in the best light-houses are made of sheet-copper, plated in the proportion of six ounces of silver to sixteen ounces of copper. They are moulded to the parabaloïdal form by a delicate and laborious process of beating with mallets and hammers of various forms and materials, and are frequently tested during the operation by the application of a mould carefully formed. After being brought to the curve, they are stiffened round the edge by means of a strong bizzle, and a strap of brass which is attached to it for the purpose of preventing any accidental alteration of the figure of the reflector. Polishing powders are then applied, and the instrument receives its last finish.　　*　　*　　*　　*

In light-houses of moderate height, the proper position for the reflector itself is perfect horizontality of its axis, which may be ascertained with sufficient accuracy by trying with a plummet whether the lips of the instrument, which we may conclude to be at right angles to the plane of its axis, be truly vertical. In light-rooms very much elevated above the sea, however, the dip of the horizon becomes notable; and a slight inclination forwards should be given to the face of the reflectors, so that their axes produced may be tangents to the earth at the visible horizon of the light-room; an arrangement which, in practice, may be easily made by reflecting the sea horizon in a small mirror placed at the focus, and inclined at 45° to the axis of the paraboloïd, so that the image of the sea-line may reach the eye in the line of the parameter in the same manner, as is afterwards noticed in speaking of the inclination of the curved mirrors used in addition to the refractors in certain dioptric fixed lights. This dip of the reflector, however, must not be permitted to interfere with the perfect horizontality of the top of the burner, which is indispensable to its proper burning.

*   *   *   *   *   *   *   *

The effect of an annular lens, in combination with the great lamp, may be estimated at moderate distances to be nearly equal to that of between 3,000 and 4,000 Argand flames of about an inch diameter; that of a cylindric refractor at about 250; and that of a curved mirror may perhaps on an average be assumed at about 10 Argand flames.

The dioptric lights used in France are divided into six orders, in relation to their power and range; but in regard to their characteristic appearances, this division does not apply, as, in each of the orders, lights of identically the same character may be found, differing only in the distance at which they can be seen, and in the expense of their maintenance. The six orders may be briefly described as follows :

1. Lights of the first order having an interior radius or focal distance of 36.22 inches (92$^{cm.}$), and lighted by a lamp of four concentric wicks, consuming five hundred and seventy gallons of oil per annum.

2. Lights of the second order, having an interior radius of 27.55 inches (70$^{cm.}$), lighted by a lamp of three concentric wicks, consuming three hundred and eighty-four gallons of oil per annum.

3. Lights of the third order, lighted by a lamp of two concentric wicks, consuming one hundred and eighty-three gallons of oil por annum, and having a focal distance of 19.68 inches (50$^{cm.}$)

4. Lights of the fourth order, or harbor-lights, having an internal radius of 9.84 inches (25$^{cm.}$), and a lamp of two concentric wicks, consuming about one hundred and thirty gallons of oil per annum.

5. Lights of the fifth order, having a focal distance of 7.28 inches (18.5$^{cm.}$); and

6. Lights of the sixth order, having an internal radius of 5.9 inches (15$^{cm.}$), and lighted by a lamp of one wick, or Argand burner, consuming forty-eight gallons of oil per annum, The more minute subdivisions of orders I consider to be unnecessary.

Those orders are not intended as distinctions, but are characteristic of the power and range of lights, which render them suitable
for different localities on the coast, according to the distance at
which they can be seen. This division, therefore, is analogous to
that which separates our lights into *sea lights, secondary lights,* and
harbor lights, terms which are used to designate the power and
position, and not the appearance of the lights to which they are
applied.

Each of the above orders is susceptible of certain combinations,
which produce various appearances, and constitute the distinctions
used for dioptric lights; but the following are those which have been
actually employed as the most useful in practice :

The first order contains, first, lights producing, once in every
minute, a great flash, preceded by a smaller one, by the revolution
of eight great lenses and eight smaller ones combined with eight
mirrors; second, lights flashing once in every half minute, and composed of sixteen half lenses. Those lights may have the subsidiary
parts simply catoptric, or diacatoptric ; and, third, fixed lights,
composed of a combination of cylindric pieces, with curved mirrors
or catadioptric zones ranged in tiers above and below them.

The second order comprises revolving lights with sixteen or twelve
lenses, which make flashes every half minute; and fixed lights varied
by flashes once in every four minutes—an effect which, as already
noticed, is produced by the revolution of exterior cylindric pieces.

The third order contains common fixed lights, and fixed lights
varied by flashes once in every four minutes.

The fourth order contains simple fixed lights, and fixed lights varied
by flashes once in three minutes.

The fifth order has fixed lights varied by flashes once in every
three minutes, and fixed lights of the common kind. It has been
thought necessary to change the term "fixed lights varied by
flashes" for "fixed lights with short eclipses," because it has been
found that at certain distances a momentary eclipse precedes the
flash. The sixth order has only fixed lights.

These distinctions depend upon the periods of revolution rather
than upon the *characteristic appearance* of the light ; and therefore
seem less calculated to strike the eye of a seaman than those employed on the coasts of Great Britain and Ireland. In conformity

with this system, and in consideration of the great loss of light which results from the application of colored media, distinctions based upon color have been generally discarded in the French lights.

The distinctions are in fact only *four* in number, viz: fixed; fixed, varied by flashes ;* revolving, with flashes once a minute ; and revolving with flashes every half minute. To those might be added, revolving, with bright periods once in two minutes, and perhaps *flashing* once in *five seconds*, (as introduced by me at the Little Ross, but I cannot say with such complete success as would induce me to recommend its general adoption.) My own experience would also lead me to reject the distinction called "fixed, varied by flashes," which I do not consider as possessing a marked or efficient character.

Having thus fully described the nature of the catoptric and dioptric modes of illuminating light-houses, I shall conclude with a comparative view of the merits of both systems, deduced from the experiments made at Gullan-hill during the winters of 1832 and 1833, under the inspection of the commissioners of Northern lights. The chief practical result of those trials was, that the light of one of the great annular lenses used in the revolving lights of the first order, was equal to the united effect of eight of the large reflectors employed in the revolving lights on the Scotch coast. It may be said, however, that the diacatoptric† combination of pyramidal lenses and plane mirrors of Cordouan, adds the power of more than two reflectors to the effect of the great lens ; but it ought to be remembered that in the French lights this additional power is used only to compensate for one of the defects of the system by lengthening the duration of the flash, and therefore contributes, if at all, only in a very indirect manner to render the light visible to the mariner at a greater distance. M. Fresnel found that from the smaller divergence of the lens the eclipses were too long and the bright periods of the revolution too short, and he therefore determined to adopt the horizontal deviation of 7° for the upper lenses, with a view to remedy this defect. Assuming, therefore, that it were required to increase the number of reflectors in a revolving light of three sides, so as to render it equal in power to a dioptric revolving light of the first order, it would be necessary to place eight reflectors on each face, so that the greatest number of reflectors required for this

---

º The "Feu fixe, varié par des éclats," or "Feu fixe, a courtes éclipses," of Fresnel.

† I use this word to designate the arrangement of pyramidal lenses and plane mirrors, by which the light is first *refracted* and then *reflected*.

purpose may be taken at *twenty-four*. M. Fresnel has stated the expenditure of oil in the lamp of four concentric wicks at seven hundred and fifty grammes of colza oil per hour ; and it is found by experience at the Isle of May and Inchkeith that the quantity of spermaceti oil consumed by the great lamp is equal to that burned by from fourteen to sixteen of the Argand lamps used in the Scotch lights. It therefore follows that by dioptric means the consumption of oil necessary for between fourteen and sixteen reflectors will produce a light as powerful as that which would require the oil of twenty-four reflectors in the catoptric system of Scotland ; and consequently that there is an excess of oil equal to that consumed by ten reflectors, or four hundred gallons in the year, against the Scotch system. But in order fully to compare the economy of producing two revolving lights of equal power by those two methods, it will be necessary to take into the calculation the interest of the first outlay in establishing them.

The expense of fitting up a revolving light with twenty-four reflectors, ranged on three faces, may be estimated at £1,298, and the annual maintenance, including the interest of the first c st of the apparatus, may be calculated at £418 8s. 4d. The fitting up a revolving light with eight lenses and the diacatoptric accessory apparatus, may be estimated at £1,459, and the anuul maintenance at £354 10s. 4d. It therefore follows that to establish and afterwards maintain a catoptric light of the kind called *revolving white*, with a frame of three faces, each equal in power to a face of the dioptric light of Cordouan, an annual outlay of £63 18s. more would be required for the reflecting light than for the lens light ; while for a light of the kind called *revolving red and white*, whose frame has four faces, at least thirty-six reflectors would be required in order to make the light even approach an equality to that of Cordouan ; and the catoptric light would in that case cost £225 more than the dioptric light.

The effect produced by burning an equal quantity of oil in revolving lights on either system may be estimated as follows : In a revolving light like that of Skerryvore, having eight sides, each lighting with its greatest power a horizontal sector of 4°, we have 32° (or *units*) of the horizon illuminated with the full power of three thousand two hundred Argand flames, and consequently an aggregate effect of one hundred and two thousand four hundred flames, produced by burning the oil required for *sixteen* reflectors ; while in a catoptric apparatus, like that of the old light at Inchkeith, having seven sides of one reflector, each lighting with its greatest power a sector of 4° 25', we have nearly 31° (or *units*) of the horizon illuminated with the full power of four hundred Argand flames, and consequently an aggregate effect of twelve thousand four hundred flames as the result of burning the oil required for *seven* reflectors. Hence the *effect* of burning the same quantity of oil in revolving lights on either system will be represented respectively by $\frac{16}{7}$ 12,400 = 28,343 for the catoptric, contrasted with 102,400 for the dioptric light; or, in other words, revolving lights on the dioptric principle use the oil more *economically* than those on the catoptric plan, nearly in the ratio of 3.6 to 1.

Let us now speak of fixed lights, to which the dioptric method is peculiarly well adapted. The effect produced by the consumption of a gallon of oil in a fixed light, with twenty-six reflectors, which is the smallest number that can be properly employed, may be estimated as follows : The *mean* effect of the light spread over the horizontal sector, subtended by one reflector, as deduced from measurements made at each horizontal degree, by the method of shadows, is equal to 174 unassisted Argand burners. If, then, this quantity be muliplied by 360 degrees, we shall obtain an aggregate effect of 62,640, which, divided by 1,040, (the number of gallons burned during a year in *twenty-six* reflectors,) would give sixty Argand flames for the effect of the light maintained throughout the year by the combustion of a gallon of oil. On the other hand, the power of a catadioptric light of the first order, like that lately established at Girdleness, may be estimated thus : The *mean* effect of the light produced by the joint effect of both the dioptric and catadioptric parts of fixed light apparatus, may be valued at 450 Argand flames, which, multiplied by 360 degrees, gives an aggregate of 162,000 ; and if this quantity be divided by 570 (the number of gallons burned by the great flame in a year) we shall have about 284 Argand flames for the effect of the light produced by the combustion of a gallon of oil. It would thus appear that in fixed lights, the French apparatus, as lately improved, produces, as the *average* effect of the combustion of the same quantity of oil over the whole horizon, upwards of *four times* the amount of light that is obtained by the catoptric mode.

But the great superiority of the dioptric method chiefly rests upon its *perfect* fulfilment of an important condition required in a fixed light, by distributing the rays *equally* in every point of the horizon. In the event of the whole horizon not requiring to be illuminated, the dioptric light would lose a part of its superiority in economy, and when half the horizon only is lighted, it would be more expensive than the reflected light ; but the greater power and more equal distribution of the light may be considered of so great importance, as far to outweigh the difference of expense. In the latter case, too, an additional power, as already noticed, can be given to the dioptric light, by placing at the landward side of the light room, spherical mirrors with their centres in the focus of the refracting apparatus.* The luminous cones, or pyramids, of which such reflectors would form the bases, instead of passing off uselessly to the land, would thus be thrown back through the focal point, and finally refracted, so as to increase the effect of the light seaward by nearly *one-third* of the light which would otherwise be lost.

The expense of establishing a fixed light composed of twenty-six reflectors may be estimated at £950, and its annual maintenance, including interest on the first cost of the apparatus, may be reckoned at £425 10s.; and the expense of fitting up a fixed light on the dioptric principle with catadioptric zones is £1,511, while its annual maintenance may be taken at £285 6s. 4d. It thus appears that the annual expenditure of dioptric fixed light is £140 3s. 8d. less than that of a fixed light composed of twenty-six reflectors; while the *average* effect, equally diffused over the horizon, is *four times* greater.

The comparative views already given of the catoptric and dioptric modes of illuminating light-houses, demonstrate that the latter produces more powerful lights by the combustion of the same quantity of oil : while it is obvious that the catoptric system insures a more certain exhibition of the light, from the fountain lamps being less liable to derangement than the mechanical lamps used in dioptric lights. The balance, therefore, of real advantages or disadvantages, and consequently the propriety of adopting the one or the other system, involves a mixed question, not susceptible of a very precise solution, and leaving room for different decisions, according to the

---

° A similar arrangement can also be made in revolving lights by making the radius of the mirrors somewhat less than that of the inscribed circle of the octagon bounded by the lenses, so that they may circulate freely round the backs of the mirrors. The shortness of the radius of the reflecting surface would, of course, increase the divergence of the beam of light refracted through the lenses, as the flame would, in this case, subtend a greater angle at the face of the mirrors.

value which may be set upon obtaining a cheaper and better light on the one hand, as contrasted, on the other, with less certainty in its exhibition. Experience, however, goes far to show that, in practice, the risk of extinction of the lamp in dioptric lights is very small.

A few general considerations, serving briefly to recapitulate the arguments for and against the two systems, may not be out of place. And, first, regarding the fitness of dioptric instruments for revolving lights, it appears from the details above given—

1. That by placing eight reflectors on each face of a revolving frame, a light might be obtained as brilliant as that derived from the great annular lens; and that, in the case of a frame of three sides, the excess of expense by the reflecting mode would be £63 18s.; and in the case of a frame of four sides, the excess would amount to £225.

2. That for burning oil economically in revolving light-houses, which illuminate every point of the horizon successively, the lens is more advantageous than the reflector in the ratio of 3.6 to 1.

3. That the divergence of the rays from the lens being less than from the reflector, it becomes difficult to produce, by lenses, the appearance which characterizes the catoptric revolving lights, already so well known to British mariners; and any change of existing lights which would, of course, affect their appearance, must, therefore, involve some practical objections, which do not at all apply to the case of new lights.

4. That the uncertainty in the management of the lamp renders it more difficult to maintain the revolving dioptric lights without risk of extinction—an accident which has several times occurred at Cordouan and other light-houses, both in France and elsewhere. A more extended experience, however, has tended to moderate any fears on this head.

5. That the extinction of one lamp in a revolving catoptric light is not only less probable, but leads to much less serious consequences than the extinction of the single lamp in a dioptric light; because, in the first case, the evil is limited to diminishing the power of *one face* by an *eighth* part; whilst, in the second, the *whole horizon is totally deprived of light*. The extinction of a lamp, therefore, in a dioptric light, leads to evils which may be considered *very great* in comparison with the consequences which attend the same accident in a catoptric light.

In comparing the fixed dioptric and the fixed catoptric apparatus, the results may be summed up under the following heads :

1. It is impossible, by means of any practicable combination of paraboloïdal reflectors, to distribute round the horizon a zone of light of exactly *equal intensity ;* while this may be easily effected by dioptric means, in the manner already described. In other words, the qualities required in fixed lights cannot be so fully obtained by reflectors as by refractors.

2. The *average* light produced in every azimuth by burning one gallon of oil in Argand lamps, with reflectors, is only about *one-fourth* of that produced by burning the same quantity in the dioptric apparatus ; and the annual expenditure is £140 3s. 8d. less for the entire dioptric light than for the catoptric light.

3. The *characteristic* appearance of the fixed reflecting light in any one azimuth would not be changed by the adoption of the dioptric method, although its increased *mean* power would render it visible at a greater distance in every direction.

4. From the equal distribution of the rays, the dioptric light would be observed at equal distances in every point of the horizon; an effect which cannot be fully attained by any practicable combination of paraboloïdal reflectors.

5. The inconveniences arising from the uncertainty which attends the use of the mechanical lamp, are not perhaps so much felt in a fixed as in a revolving light; because the greater simplicity of the apparatus admits of easier access to it in case of accident.

6. But the extinction of a lamp in a catoptric light, leaves only one *twenty-sixth part* of the horizon without the benefit of the light, and the chance of accident arising to vessels from it, may, therefore, be considered as incalculably less than the danger resulting from the extinction of the single lamp of the dioptric light, which deprives the whole horizon of light.

7. There is also, in certain situations, a risk arising from irregularity in the distances at which the same fixed catoptric light can be seen in the different azimuths. This defect, of course, does not exist in the dioptric light.

There can be little doubt that the more fully the system of Fresnel is understood, the more certainly will it be preferred to the catoptric system of illuminating light-houses, at least in those countries where this important branch of administration is conducted with the care and solicitude which it deserves.

It must not, however, be imagined, that there are no circumstances in which the catoptric system is not absolutely preferable to illumination by means of lenses. We have hitherto attended only to horizontal divergence and its effects, and this is unquestionably the more important view; but the consideration of vertical divergence must not be altogether overlooked. Now while it is obvious that vertical divergence, at least above the horizon, involves a total loss of the light which escapes uselessly upwards into space, (in which respect the reflectors are much less advantageous,) it is no less true, that if the sheet of light which reaches the most distant horizon of the light-house, however brilliant, were as *thin* as the absence of all vertical divergence would imply, it would be practically useless; and some measure of dispersion in the arc *below* the horizon is therefore absolutely indispensable to constitue a really useful light. In the reflector, the greatest vertical divergence below the horizontal plane of the focus is 16° 8′, and that of the lens is about 4° 30′. Let us consider for a moment the bearing of those facts upon the application of the two modes of illumination to special circumstances. The powerful beam of light transmitted by the lens peculiarly fits that instrument for the great sea-lights which are intended to warn the mariner of his approach to a distant coast which he first makes on an *over-sea* voyage; and the deficiency of its divergence, whether horizontal or vertical, is not practically felt as an inconvenience in lights of that character, which seldom require to serve the double purpose of being visible at a great distance, and at the same time of acting as guides for danger near the shore. For such purposes, the lens applies the light much more advantageously as well as more economically than the reflector; because, while the duration of its *least* divergent beam is nearly equal to that of the reflector, it is *eight* times more powerful. A revolving system of eight lenses illuminates an horizontal arc of 32° with this bright beam. The reflector, on the other hand, spreads the light over a larger arc of the horizon; and, while its *least* divergent beam is much less powerful than that of the lens, the light which is shed over its *extreme* arc is so feeble as to be practically of no use in lights of extensive range, even during clear weather. When a light-house is placed on a very high headland, however, the deficiency of divergence in the vertical direction is often found to be productive of some practical inconvenience; but this defect may be partially remedied by giving to the lenses a slight inclination *outwards* from the vertical plane of the focus, so as to cause the most brilliant portion of the emergent beam to reach the *visible horizon* which is due to the height of the lantern. It may be observed, also, that a lantern at the height of 150 feet, which (taking into account the common height of the observer's eye at sea) commands a range of upwards of twenty English miles, is sufficient for all the ordinary purposes of the navigator, and that the intermediate space is practically easily illuminated, even to within a mile of the light-house, by means of a slight inclination of the subsidiary mirrors, even where the light from the principal part of the apparatus passes over the seaman's head. For the purpose of leading lights, in narrow channels, on the other hand, and for the illumination of certain narrow seas, there can be no doubt that reflectors are much more suitable and convenient. In such cases, the amount of vertical divergence below the horizon forms an important element in the question, because it is absolutely necessary that the mariner should keep sight of the lights even when he is very near them; while there is not the same call for a very powerful beam which exists in the case of sea-lights. Yet even in narrow seas, where low towers, corresponding to the extent of the *range* of the light, are adopted, but where it is, at the same time, needful to illuminate the whole or the greater part of the horizon, the use of dioptric instruments will be found almost unavoidable, especially in fixed lights, as well from their equalizing the distribution of the light in every azimuth, as from their much greater economy in situations where a large annual expenditure would often be disproportionate to the revenue at disposal. In such places, where certain peculiarities of the situation require the combination of a light equally diffused over the greater portion of the horizon, along with a greater vertical divergence in certain azimuths, than dioptric instruments afford, I have found it convenient and economical to add to the fixed refracting apparatus a single paraboloïdal reflector in order to produce the desired effect, instead of adapting the whole to the more expensive plan for the sake of meeting the wants of a single narrow sector of its range. In other cases, where the whole horizon is to be illuminated, and great vertical divergence is at the same time desirable, a slight elevation of the burner, at the expense, no doubt, of a small loss of light, is sometimes resorted to, and is found to produce, with good effect, the requisite depression of the emergent rays.

In certain situations, where a great range, and, consequently, a powerful light must be combined with tolerably powerful illumination in the immediate vicinity of the light-house, we might, perhaps, advantageously adopt a variation of the form and dimensions of the mirrors employed, so as to resemble those formerly used at the Tour de Cordouan, which were of considerably larger surface and longer focal distance than those which are used in Britain. If such a form were adopted, the power of the light for the purpose of the distant range would be increased ; and I would propose to compensate for the deficiency of divergence consequent on a long focal distance, by placing a second burner in some position between the parameter and the vertex, and slightly elevated above the axis of the instrument, so as to throw the greater portion of the beam resulting from this second burner below the horizontal plane of the focus. Such an expedient is no doubt somewhat clumsy, and would at the same time involve the consumption of twice the quantity of oil used in an ordinary catoptric light; but I can still conceive it to be preferable, in certain situations, to the use of the lenses alone.

Thus it appears that we must not too absolutely conclude against one, or in favor of the other mode of illumination for light-houses; but, as in every other department of the arts, we shall find the necessity of patiently weighing all the circumstances of each particular case that comes before us, before selecting that instrument, or combination of instruments, which appears most suitable.

The mode of distinguishing lights in the system of Fresnel depends more upon their *magnitude* and the *measured interval* of the time of their revolution, than upon their *appearance;* and no other very marked distinctions, except fixed and revolving, have been successfully attempted in France. As above stated, I consider the distinction of the *fixed light varied by flashes*, to possess an appearance too slightly differing from that of a revolving light, to admit of its being safely adopted in situations where revolving lights are near. The trial which I made at the Little Ross, in the Solway Frith, of producing, by means of lenses, a light flashing once in five seconds of time, although successful so far as mere distinction is concerned, has several practical defects, arising from the shortness of the duration of the flashes compared with the powerful effect of the fixed part of the apparatus, which I consider sufficient to prevent its adoption in future, especially considering that a much more marked appearance can be produced by means of reflectors, as has been done at the Buchanness in Aberdeenshire, and the Rhinns of Islay in Argyleshire. Colored media have never, so far as I know, been applied to dioptric apparatus, except in the case of the Maplin light at the mouth of the Thames, and Cromarty Point light at the entrance to the Cromarty Frith, Nosshead in Caithness, and Ship Rock of Sanda in Argyleshire, but in all those instances successfully. In the case of the fixed light at Sanda, in particular, I would observe that it is seen at the distance of sixteen nautical miles, and occasionally observed even so far off as twenty-two nautical miles. The enormous loss of light, however, amounting to no less than 0·80 of the whole incident rays, forms a great bar to the adoption of color as a distinction; and any means which could tend to lessen that absorption, and at the same time produce the characteristic appearance, would be most valuable. I have tried some glasses of a pink tinge, prepared by M. Letourneau of Paris, in which the absorption does not exceed 0·57 of the incident rays; but the appearance of the light, at a distance, is much less marked than that produced by the glasses used in Britain. Such deficiency of characteristic color might lead to serious consequences, as the transmission of white rays, through a hazy atmosphere, too often produces, by absorption, a reddish tinge of the light, for which the less marked appearance given by the paler media might be easily mistaken. This coloring power of absorption is so well known, that red lights are seldom used except in direct contrast with white ones; but, on a coast so thickly studded with light-houses as that of Great Britain, the number of distinctions is insufficient to supply all our wants, so that we are sometimes reluctantly compelled to adopt a *single red light* in some situation of lesser importance, or which, from some local circumstances and the appearance of the lights which must be seen by the mariner before passing it, is not likely to be mistaken for any other. The great loss of light by colored media causes the red beam, in a revolving light, to be seen at a shorter distance than the white; and it is conceivable that, in certain circumstances, this might lead the mariner to mistake a *red and white light* for a *white* light revolving at half the velocity. Such a mistake might perhaps prove dangerous; but the lights are generally so situated that there is ample time for the mariner, after first discovering the red light, and thus correcting any mistake, to shape his course accordingly. All other colored media, except *red*, have been found useless as distinctions for any lights of extensive range, and fail to be efficient, owing to the necessity of absorbing almost all the light before a marked appearance can be obtained. In a few pier or ferry lights, green and blue media have been tried, and found available at the distance of a few cables' lengths.*

It seems to be a natural consequence of the physical distribution of light, that fixed lights, which illuminate the whole horizon, should be less powerful than revolving lights which have their effect concentrated within narrow sectors of the horizon. Any attempt to increase the power of fixed lights is, therefore, worthy of attention; and when the late Captain Basil Hall proposed a plan for effecting this object, it received, as it deserved, the full consideration of the Scotch Light-house Board, who authorized me to repeat Captain Hall's experiments, and verify his results by observations made at a considerable distance.

The familiar experiment of whirling a burning stick quickly round the head, so as to produce a ribbon of light, proves the possibility of causing a continuous impression on the retina by intermittent images succeeding each other with a certain rapidity. From the moderate velocity at which this continuity of impression is obtained, we should be warranted in concluding, *a priori*, that the time required to make an impression on the retina is considerably less than the duration of the impression itself; for the continuity of effect must, of course, be caused by fresh impulses succeeding each other before the preceding ones have entirely faded. If it were otherwise, and the time required to make the impression were equal to the duration of the sensation, it would obviously be impossible to obtain a series of impulses so close or continuous in their effect as to run into and overlap each other, and thus throw out the intervals of darkness; because the same velocity which would tend to shorten the dark intervals, would also curtail the bright flashes, and thus prevent their acting on the eye long enough to cause an impression. Accordingly, we find that the duration of an impression is in reality much greater than the time required for producing the effect on the retina. It is stated by Professor Wheatstone, in the London Transactions for 1834, that only about *one millionth part* of a second is required for making a distinct impression on the eye; and it appears from a statement made by Lamé, at p. 425 of his Cours de Physique, that M. Plateau found that an impression on the retina preserves its intensity unabated during *one*

---

* In some late experiments which I made with very powerful instruments, green lights were visible, in very clear weather, at the distance of seven miles. The blue could only once be seen, with great difficulty, at five miles.

hundredth of a second, so that, however small those times may be in themselves, the one is ten thousand times greater than the other.

It has been ascertained, by direct experiment,[*] that the eye can receive a fresh impression before the preceding one has faded ; and, indeed, if this were impossible, absolute continuity of impression from any succession of impulses, however rapid, would seem to be unattainable; and the approach to perfect continuity would be inversely at the time required to make an impression.

From the property which bright bodies passing rapidly before the eye possess of communicating a continuous impression to the sense of sight, the late Captain Basil Hall conceived the idea, not merely of obtaining all the effects of a fixed light, by causing a system of lenses to revolve with such a velocity as to produce a continuous impression, but, at the same time, of obtaining a much more brilliant appearance, by the compensating influence of the bright flashes, which he expected would produce impulses sufficiently powerful and durable to make the deficiency of light in the dark spaces almost imperceptible. The mean effect of the whole series of changes would, he imagined, be thus greatly superior to that which can be obtained from the same quantity of light equally distributed, as in fixed lights, over the whole horizon. Now this expectation, if it be considered solely in reference to the physical distribution of the light, involves various difficulties. The quantity of light subjected to instrumental action is the same whether we employ the refracting zones at present used in fixed dioptric lights, or attempt to obtain continuity of effect by the rapid revolution of lenses ; and the only difference in the action of those two arrangements is this, that while the zones distribute the light equally over the whole horizon, or rather do not interfere with its natural horizontal distribution, the effect of the proposed method is to collect the light into pencils, which are made to revolve with such rapidity that the impression from each pencil succeeds the preceding one in time to prevent a sensible occurrence of darkness. To expect that the mean effect of the light, so applied, should be greater than when it is left to its natural horizontal divergence, certainly appears at first to involve something approaching to a contradiction of physical laws. In both

[*] Lamé, Cours de Physique, p. 424. "L'impression peut subsister encore lorsque la suivante a lieu."

cases, the same quantity of light is acted upon by the instrument; and, in either case, any one observer will receive an impression similar and equal to that received by any other stationed at a different part of the horizon; so that, unless we imagine that there is some loss of light peculiar to one of the methods, we are shut up, in the physical view of the question, to the conclusion that the impressions received by each class of observers must be of equal intensity. In other words, the same quantity of light is by both methods employed to convey a continuous impression to the senses of spectators in every direction, and, in both methods, equality of distribution is effected, since it does not at all consist with our hypothesis, that any one observer in the same class should receive more or less than his equal share of the light. Then, as to the probability of the loss of light, it seems natural to expect that this should occur in connection with the revolving system, because the velocity is an extraneous circumstance, by no means necessary to an equal distribution of the light, which can, as we already know, be more naturally, and at the same time perfectly attained by the use of the zones.

On the other hand, it must not be forgotten that, although the effect of both methods is to give each part of the horizon an equal share of light, there is yet this difference between them, that while the light from the zones is equally intense at every instant of time, that evolved by the rapidly circulating lenses is constantly passing through every phase between total darkness and the brightest flash of the lens; and this difference, taken in connection with some curious physiological observations regarding the sensibility of the retina, gives considerable countenance to the expectation on which Captain Hall's ingenious expedient is based. The fact which has already been noticed, and which the beautiful experiments of M. Plateau and Professor Wheatstone have of late rendered more precise, that the duration of an impression on the retina is not only appreciable, but is much greater than the time required to cause it, seems to encourage us in expecting that, while the velocity required to produce continuity of effect would not be found so great as to interfere with the formation of a full impression, the duration of the impulse from each flash would remain unaltered, and the dark intervals which do not excite the retina would, at the same time, be shortened; and that, therefore, we might, in this manner, obtain an effect on the

senses exceeding the brilliancy of a steady light distributed equally
in every direction by the ordinary method. Some persons, indeed,
who have speculated on this subject, seem even to be of opinion,
that so far from the whole effect of the series of continuous impressions being weakened by a blending of the dark with the bright
intervals, the eye would in reality be stimulated by the contrast of
light and darkness, so as thereby to receive a more complete and
durable impulse from the light. It is obvious, however, that this
question regarding the probable effect to be anticipated from a revolution so rapid as to cause a continuous impression, could only have
been satisfactorily answered by appeal to experiment.

In experimenting on this subject, I used the apparatus formerly
employed by Captain Hall. It consisted of an octagonal frame,
which carried eight of the discs that compose the central part of
Fresnel's compound lens, and was susceptible of being revolved
slowly or quickly at pleasure, by means of a crank handle and some
intermediate gearing. The experiments were nearly identical with
those made by Captain Hall, who contrasted the effect of a single
lens at rest, or moving very slowly, with that produced by the
eight lenses, revolving with such velocity as to cause an apparently
continuous impression on the eye. To this experiment I added that
of comparing the beam thrown out by the central portion of a cylindric refractor, such as is used at the fixed light of the Isle of May,
with the continuous impression obtained by the rapid revolution of
the lenses. Captain Hall made all his comparisons at the short distance of one hundred yards; and, in order to obtain some measure of
the intensity, he viewed the lights through plates of colored glass
until the luminous discs became invisible to the eye. I repeated
those experiments at Gullan, under similar circumstances, but with
very different results. I shall not, however, enter upon the discussion of those differences here, although they are susceptible of
explanation, and are corroborative of the conclusions at which I
arrived by comparing the lights from a distance of fourteen miles,
but shall briefly notice the more important results which were obtained by the distant view. They are as follows:

1. The flash of the lens revolving slowly was very much larger
than that of the rapidly revolving series; and this decrease of size
in the luminous object presented to the eye became more marked as

the rate of revolution was accelerated, so that, at the velocity of eight or ten flashes in a second the naked eye could hardly detect it, and only few of the observers saw it, while the steady light from the fixed refractor was distinctly visible.

2. There was also a marked falling off in the brilliancy of the rapid flashes as compared with that of the slow ones ; but this effect was by no means so striking as the decrease of volume.

3. Continuity of impression was not attained at the rate of five flashes in a second, but each flash appeared to be distinctly separated by an interval of darkness ; and even when the nearest approach to continuity was made, by the recurrence of eight or ten flashes in a second, the light still presented a twinkling appearance, which was well contrasted with the steady and unchanging effect of the cylindric refractor.

4. The light of the cylindric refractor was, as already stated, steady and unchanging, and of much larger volume than the rapidly revolving flashes. It did not, however, appear so brilliant as the flashes of the quickly revolving lenses, more especially at the lower rate of five flashes in a second.

5. When viewed through a telescope, the difference of volume between the light of a cylindric refractor and that produced by the lenses at their greatest velocity was very striking. The former presented a large diffuse object of inferior brilliancy, while the latter exhibited a sharp pin-point of brilliant light.

Upon a careful consideration of these facts, it appears warrantable to draw the following general conclusions :

1. That our expectations as to the effects of light, when distributed according to the law of its natural horizontal divergence, are supported by observed facts as to the visibility of such lights, contrasted with those whose continuity of effect is produced by collecting the whole light into bright pencils, and causing them to revolve with great velocity.

2. It appears that this deficiency of visibility seems to be chiefly due to a want of volume in the luminous object, and also, although in a less degree, to a loss of intensity, both of which defects appear to increase in proportion as the motion of the luminous object is accelerated.

3. That this deficiency of volume is the most remarkable optical

14

phenomenon connected with the rapid motion of luminous bodies, and that it appears to be directly proportional to the velocity of their passage over the eye.

4. That there is reason to suspect that the visibility of distant light depends on the volume of the impression in a greater degree than has perhaps been generally imagined.

5. That, as the size and intensity of the radiants causing these various impressions to a distant observer were the same, the volume of the light, and, consequently, *cæteris paribus*, its visibility, are, within certain limits, proportionate to the time during which the object is present to the eye.

Such appear to be the general conclusions which those experiments warrant us in drawing ; and the practical result, in so far as light-houses are concerned, is sufficient to discourage us from attempting to improve the visibility of fixed lights in the manner proposed by Capt. Hall, even supposing the practical difficulties connected with the great centrifugal force generated by the rapid revolution of the lenses to be less than they really are.

The decrease in the volume of the luminous object caused by the rapid motion of the lights is interesting from its apparent connection with the curious phenomenon of irradiation. When luminous bodies, such as the lights of distant lamps, are seen by night, they appear much larger than they would do by day ; and this effect is said to be produced by irradiation. M. Plateau, in his elaborate essay on this subject, after a careful examination of all the theories of irradiation, states it to be his opinion, that the most probable mode of accounting for the various observed phenomena of irradiation is to suppose, that, in the case of a night-view, the excitement caused by light is propagated over the retina beyond the limits of the day-image of the object, owing to the increased stimulus produced by the contrast of light and darkness ; and he also lays it down as a law, confirmed by numerous experiments, that irradiation increases with the duration of the observation. It appears, therefore, not unreasonable to conjecture, that the deficiency of volume observed during the rapid revolution of the lenses may have been caused by the light being present to the eye so short a time, that the retina was not stimulated in a degree sufficient to produce the amount of irradiation required for causing a large visual object.

When, indeed, the statement of M. Plateau, that irradiation is proportional to the duration of the observation, is taken in connection with the observed fact, that the volume of the light decreased as the motion of the lenses was accelerated, it seems almost impossible to avoid connecting together the two phenomena as cause and effect.

Before leaving this part of the subject, I will call attention to some late plans for combining dioptric and catoptric apparatus, the object of which is to subject to the corrective action of instruments, a greater proportion of the luminous sphere than it has yet been found practicable to do, especially in revolving lights. Reflectors act chiefly on the *posterior* portion of the flame, and generally receive about *twelve-seventeenths* of the whole luminous sphere ; while a series of dioptric instruments can only affect an *anterior zone*, amounting to about two-fifths of the whole light which is emitted by the lamp. Certain deductions due to the form of the lower part of the burners, and to the loss of light at reflection, which is not less than *one-half* of the incident light, as well as to that by refraction through the lens, (which, however, cannot exceed *one-tenth* of the incident light,) will reduce those numbers from twelve-seventeenths to one-third, and from two-fifths to three-tenths, thus making the ratio of the proportion of the whole flame *actually given forth* by the reflectors to the amount by lenses equal to that of ten to nine. In fixed lights, on Fresnel's system, we have already seen that nearly the whole of the available light is turned to a useful purpose by means of the curved mirrors or catadioptric zones, which are added to increase the effect of the central dioptric belt ; and, in revolving lights, an approximation to a similar result is obtained by the addition of the diacatoptric combination of pyramidal lenses and plane mirrors placed above the great lenses. Catoptric lights, however, to which such auxiliary arrangements are inapplicable, had still the great disadvantage of leaving the *anterior cone* of light to pass off in the useless state of naturally divergent light ; and anything calculated to increase the power of that class of lights, without altering that simplicity and security of the burners employed in them, which renders them so suitable for remote situations in the colonies, deserves careful attention. It will be remembered that the proposal of Mr. Barlow for effecting this object, has already been noticed; and it is needless now to do more than remind the reader that the practical disadvantage of the great aberration in the path of the rays reflected from the subsidiary hemispherical mirror, which must necessarily be of very small dimensions, together with the great loss of light by the second reflection, must go far to neutralize the effect of Mr. Barlow's plan. A combination of dioptric and catoptric instruments, intended to produce a similar effect, has been proposed by Mr. Alexander Gordon, and is described at page 385 of the tenth volume of the *Civil Engineers' and Architects' Journal.* It consists of a paraboloïdal mirror, of a very short focal distance, with some of the outer zones of one of Fresnel's smaller lenses in front of it. The zones are intended to refract some of the rays that escape past the edges of the mirror, while the pencil of light reflected from the mirror itself is supposed to pass through the circular space which is generally occupied by the central portion of the lens. This arrangement is a step in the right direction, only in so far as it implies the union of the two modes of illumination ; but, as it is by no means skilfully designed, it is liable to several palpable objections.

1. The actual gain of light has been greatly overrated by the writer in the Journal, who expects to turn twenty-seven-twenty-eighths of the whole light to a useful account ; but so great a gain of light can never consist with the form and position of the lower part of the flame.

2. Upwards of twenty-four-twenty-eighths of the estimated quantity would be intercepted by the paraboloïd alone, and little more than two-twenty-eights by the rings of the lens, an addition far too insignificant to warrant the adoption of so expensive an appendage to the reflector.

3. The great aberration of the rays reflected by the conoïd behind the parameter, and its small reflecting surface, must render it practically useless ; and, perhaps, nearly *one-half* of the whole light would thus be lost to the mariner. The accurate formation of a paraboloïd of such depth would also be difficult; and, considering the practical inutility of the conoïd behind the parameter, would seem to be a misapplication of labor.

4. The union of such an instrument with the lenticular zones in front, which require that the pencil of parallel rays should be reflected with the greatest accuracy, so as to enable them to pass through the circular space bounded by the zones, is an obvious misapplica-

tion of a paraboloïd with a short focal distance, to a purpose for which it is singularly unsuitable.

5. Mirrors somewhat of the same form were in use at Scilly light-house, and were long ago discarded as disadvantageous, at the suggestion of the late eminent Captain Huddart.

6. The outer zones, which form the *least efficient*, and at the same time the *most expensive* portion of the compound lens have been preferred to the central portion of that instrument; and by this means the anterior cone of rays is at the same time lost.

* * * * * * * *

A considerable practical defect in all the light-house lanterns which I have ever seen, with the exception of those recently constructed for the Scotch light-houses, consists in the vertical direction of the astragals, which of course tend to intercept the whole or a great part of the light in the azimuth which they subtend. The consideration of the improvement which I had effected in giving a diagonal direction to the joints of the fixed refractors first led me to adopt a diagonal arrangement of the framework which carries the cupola of zones, and afterwards for the astragals of the lantern. Not only is the *direction* of the astragals more advantageous for equalizing the effect of the light, but the greater stiffness and strength which such an arrangement gives to the framework of the lantern, make it safe to use more slender bars, and thus also absolutely *less* light is intercepted; the panes of glass at the same time become triangular, and are necessarily stronger than rectangular panes of equal surface. This form of lantern is extremely light and elegant, and is shown, with detailed drawings of some of its principal parts, in plate X. To avoid the necessity of painting, which, in situations so exposed as those which light-houses generally occupy, is attended with many inconveniences and no small risk, the framework of the lantern is now formed of gun-metal and the dome is of copper. A lantern for a light of the first order, twelve feet in diameter, and with glass frames ten feet high, costs, when glazed, about £1,260. In order to give the light-keepers free access to cleanse and wash the upper panes of the lantern, (an operation which in snowy weather must sometimes be frequently repeated during the night,) a narrow gangway, on which they may safely stand, is placed on the level of the top of the lower panes, and at the top of the second panes rings are provided of which the light-keepers may lay hold for security in stormy weather. A light trap-ladder is also attached to the outside of the lantern by means of which there is an easy access to the ventilator on the dome.

Great care is bestowed on the glazing of the lantern, in order that it may be quite impervious to water, even during the heavy gales. When iron is used for the frames, they are carefully and frequently painted ; but gun-metal, as just noticed, is now generally used in the Scotch light-houses. There is great risk of the glass plates being broken by the shaking of the lantern during high winds; and as much as possible to prevent this, various precautions are adopted. The arris of each plate is always carefully rounded by grinding ; and grooves about half an inch wide, capable of holding a good thickness of putty, are provided in the astragals for receiving the glass, which is quarter of an inch thick. Small pieces of lead or wood are inserted between the frames and the plates of glass against which they may press, and by which they are completely separated from the more unyielding material of which the lantern-frames are composed. Panes glazed in frames padded with cushions, and capable of being temporarily fixed in a few minutes, in the room of a broken plate, are kept ready for use in the store-room. Those framed plates are called *storm panes*, and have been found very useful on several occasions when the glass has been shattered by large sea-birds coming against it in a stormy night, or by small stones violently driven against the lantern by the force of the wind.

The ventilation of the lanterns forms a most important element in the preservation of a good and sufficient light. An ill-ventilated lantern has its sides continually covered with the water of condensation, which is produced by the contact of the ascending current of heated air; and the glass, thus obscured, obstructs the passage of the rays and diminishes the power of the light. In the northern light-houses, ventilators, capable of being opened and shut at pleasure, so as to admit from without a supply of air when required, are provided in the parapet wall on which the lantern stands ; the lantern roof also is surmounted by a cover which, while it closes the top of an open cylindric tube against the entrance of rain, and descends over it only so far as is needful for that purpose, still leaves an open air-space between it and the dome. This arrangement permits the current of heated air, which is continually flowing from the lantern through the cylindric tube, to pass between it and the outer cover, from which it finally escapes to the open air through the space between the cover and the dome. The door which communicates from the light-room through the parapet to the balcony outside, is also made the means of ventilating the light-room ; and, for that purpose, it is provided with a sliding bolt at the bottom, which, being dropped into one or other of the holes cut in the balcony for its reception, serves to keep the door open at any angle that may be found necessary. A useful precaution was introduced by my predecessor, as engineer to the Northern lights Board, in order to prevent the too rapid condensation of heated air on the large internal surface of the lantern roof, which consists in having two domes with an air-space between them. as shown in the enlarged diagrams in plate X.

An important improvement in the ventilation of light-houses was some years ago introduced by Dr. Faraday into several of the light-houses belonging to the Trinity House, and has since been adopted in all the dioptric lights belonging to the commissioners of Northern light-houses. After mentioning several proofs of extremely bad ventilation in light-houses, Dr. Faraday thus describes his apparatus :[*]

"The ventilating pipe or chimney is a copper tube, four inches in diameter; not, however, in one length, but divided into three or four pieces; the lower end of each of these pieces for about one and a half inch is opened out in a conical form, about five and a half inches in diameter at the lowest part. When the chimney is put together, the upper end of the bottom piece is inserted about half an inch into the cone of the next piece above, and fixed there by three ties or pins, so that the two pieces are firmly held together; but there is still plenty of air-way or entrance into the chimney between them. The same arrangement holds good with each succeeding piece. When the ventilating chimney is fixed in its place, it is adjused so that the lamp-chimney enters about half an inch into the lower cone, and the top of the ventilating chimney enters into the cowl or head of the lantern.

"With this arrangement, it is found that the action of the ventilating flue is to carry up every portion of the products of combustion into the cowl ; none passes by the cone apertures into the air of the lantern, but a portion of the air passes from the lantern by these apertures into the flue, and so the lantern itself is in some degree ventilated.

"The important use of these cone apertures is, that when a sudden gust or eddy of wind strikes into the cowl of the lantern it should not have any effect in disturbing or altering the flame. It is found that the wind may blow suddenly in at the cowl, and the effect never reaches the lamp. The upper, or the second, or the third, or even the fourth portion of the ventilating flue might be entirely closed, yet without altering the flame. The cone junctions in no way interfere with the tube in carrying up all the products of combustion; but if any downward current occurs, they dispose of the whole of it into the room without ever affecting the lamp. The ventilating flue is in fact a tube, which, as regards the lamp, can carry everything *up* but conveys nothing *down*."

The advantages of this arrangement, as applied to the Northern light-houses, were much less palpable than those which are described in the beginning of Dr. Faraday's paper, because their ventilation was very good before its introduction; and the flame in particular was perfectly steady, being by no means subject to derangement from sudden gusts of wind from the roof in the manner noticed above.

All the light-houses in the district of the Scotch Commissioners are under the charge of at least two light-keepers, whose duties are to cleanse and prepare the apparatus for the night illumination, to mount guard singly after the light is exhibited, and to relieve each other at stated hours, fixed by the printed regulations and instructions under which they act. The rule is, that no keeper on watch shall, under any circumstances, leave the light-room until relieved by his comrade; and, for the purpose of cutting off all pretext for the neglect of this universal law, the dwelling-houses are built close to the light-tower, and means are provided for making signals directly from the light-room to the sleeping apartments below. The signals are communicated by air tubes (Plate XII) which pass from the light-room to the sleeping apartments in the houses, and through which, by means of a small piston, or puff of wind from the mouth, calls can be exchanged between the keepers. The man on guard in the light-room, at the end of the watch or on any sudden emergency, may thus summon his comrade from below, who, on being thus called, answers by a counter-blast, to show that the summons has been heard and will be obeyed. For the purpose of greater security, in such situations as the Bell Rock and the Skerryvore, four keepers are provided for one light-room, one being always ashore on leave with his family, and the other three being at the light-house, so that in case of the illness of one light-keeper an efficient establishment of two keepers for watching the light may remain. At all the land light-houses, also, an agreement is made with some steady person residing in the neighborhood, who is instructed in the management of the light and cleansing of the apparatus, and comes under an obligation to do duty in the light-room when called upon, in the event of the sickness or absence of one of the light-keepers. This person is called the *occasional keeper*, and receives pay only while actually employed at the light-house; but in order to keep him in the practice of the duty, he is required to serve in the light-room for a fortnight annually in the month of January. For the more minute details of the light-keepers' duty, I would refer the reader to the instructions already alluded to, which will be found at the end of this volume.

Each of the two light-keepers has a house for himself and family, both being under a common roof, but entering by separate doors, as shown in Plates XI and XII, which exhibit the buildings for the new light-house at Adnarmurchan point, on the coast of Argyleshire. The principal keeper's house consists of six rooms, two of which are at the disposal of the visiting officers of the board, whose duty in inspecting the light-house or superintending repairs may call them to the station; and the assistant has four rooms, one of which is used as a barrack-room for the workmen who, under the direction of the foreman of the light-room works, execute the annual repairs of the apparatus.

The early light-houses contained accommodation for the light-keepers in the tower itself, but the dust caused by the cleaning of those rooms in the tower was found to be very injurious to the delicate apparatus and machinery in the light-room. Unless, therefore, in situations such as the Eddystone, the Bell Rock, or the Skerryvore, where it is unavoidable, the dwellings of the light-keepers ought not to be placed in the light-tower, but in an adjoining building.

Great care should be bestowed to produce the utmost cleanliness in everything connected with a light-house, the optical apparatus of which is of such a nature as to suffer materially from the effect of dust in injuring its polish. For this purpose, covered ash-pits are provided at all the dwelling-houses, in order that the dust of the fire-places may not be carried by the wind to the light-room; and, for similar reasons, iron floors are used for the light-rooms instead of stone, which is often liable to abrasion, and all the stone work near the lantern is regularly painted in oil.

If, in all that belongs to a light-house, the greatest cleanliness be desirable, it is in a still higher degree necessary in every part of the light-room apparatus, without which the optical instruments and the machinery will neither last long nor work well. Every part of the apparatus, whether lenses or reflectors, should be carefully freed from dust before being either washed or burnished; and without such a precaution, the cleansing process would only serve to scratch them. For burnishing the reflectors, prepared *rouge* (tritoxide of iron) of the finest description, which should be in the state of an impalpable powder of a deep orange-red color, is applied, by means of soft chamois skins, as occasion may require; but the great art of keeping reflectors clean consists in the daily, patient, and skilful application of manual labor in rubbing the surface of the instrument with a perfectly dry, soft and clean skin, without rouge. The form of the hollow paraboloïd is such that some practice is necessary in order to acquire a free movement of the hand in rubbing reflectors; and its attainment forms one of the principal lessons in the course of the preliminary instruction to which candidates for the situation of a light-keeper are subjected at the Bell Rock light-house. For cleansing the lenses and glass mirrors, spirits of wine is used. Having washed the surface of the instrument with a linen cloth steeped in spirits of wine, it is carefully dried with a soft and dry linen rubber, and finally rubbed with a fine chamois skin, free from any dust which would injure the polish of the glass, as well as from grease. It is sometimes necessary to use a little fine rouge with a chamois skin, for restoring any deficiency of polish which may occur from time to time; but in a well managed light-house this application will seldom, if ever, be required.

The machinery of all kinds, whether that of the mechanical lamp or the revolving apparatus, should also be kept scrupulously clean, and all the journals should be regularly and carefully oiled.   *   *

There are now no fewer than twenty-six floating lights on the coast of England.

By the kindness of the Elder Brethren of the Corporation of Trinity House of Deptford Strond, I am enabled to give the following brief sketch of the nature and peculiarities of floating lights, which was communicated to me by Mr. Herbert, the secretary of the corporation:

"The annual expense of maintaining a floating light, including the wages and victualing of the crew, who are eleven in number, is on an average £1,000; and the first cost of such a vessel, fitted complete with lantern and lighting apparatus, anchors, cables, &c., is nearly £5,000. The lanterns are octagonal in form, five feet six inches in diameter; and, where fixed lights are exhibited, they are fitted with eight Argand lamps, each in the focus of a parabolic reflector of twelve inches diameter; but, in the revolving lights, four lamps and reflectors only are fitted. The greatest depth of water in which any light-vessel belonging to the Corporation of Trinity House of Deptford Strond at present rides, is about forty fathoms, (which is at the station of the *Seven Stones* between the Scilly islands and the coast of Cornwall.)

"The corporation's light-vessels are moored with chain-cables of 1½ inch diameter, and a single mushroom anchor of 32 cwt., in which cases the chain-cables are 200 fathoms in length; some of said vessels are moored to *span-ground* moorings, consisting of 100 fathoms of chain to each arm, and a mushroom anchor of similar weight at the end of each; a riding cable of 150 fathoms being in such cases attached to the centre ring of the ground chain. The tonnage and general dimensions of the light-vessels are given on the drawing of the lines."

Still lower in the scale of "signs and marks of the sea," are beacons and buoys, which are used to point out those dangers which, either owing to the difficulty and expense that would attend the placing of more efficient marks to serve by night as well as by day, are necessarily left without lights, or which, from the peculiarity of their position, in passages too intricate for navigation by night, are, in practice, considered to be sufficiently indicated by day-marks alone. Beacons, as being more permanent, are preferred to buoys; but they are generally placed only on rocks or banks which are dry at some periods of the tide. On rocks, in exposed situations, beacons are sometimes of squared masonry, secured by numerous joggles; but, in situations difficult of access, and in which works of uncompleted masonry could not be safely left during the winter season, an open frame-work of cast-iron pipes, firmly trussed and braced, and secured to the rock with strong *louis-bats*, is preferred. The details of this frame-work are shown at plate XIII. A stone beacon, of about forty feet high, may be erected for about £700, and an iron beacon for about £640. In less exposed places, where the bottom is rock, gravel, or hard sand, a conical form of beacon, composed of cast-iron plates, united with flanges and screws, with rust joints between them, and partially filled with concrete, is sometimes used. A beacon of that kind can be erected for about £400.

Lastly, buoys, which may be regarded as the least efficient kind of mark, and as bearing the same relation to a beacon that a floating light does to a light-house, are used to mark by day *dangers* which are always covered even at low water, and also to line out the fairways of channels. They are of three kinds, viz : the *nun-buoy*, in the form of a parabolic spindle, generally truncated at one end, so as to carry a mast or frame of cage-work, and loaded at the other end, so as to float in a vertical position ; the *can-buoy*, which is a conoid floating on its side ; and, lastly, the *cask-buoy*, which is a short frustum of a spindle truncated at both ends, but almost exclusively used for carrying the warps of vessels riding at moorings. Those buoys are of various sizes and differ in cost. Mast buoys, from ten to fifteen feet in length, cost from £23 15s. to £48 ; and those of the Ribble and the Tay, which are twenty-one and twenty-four feet long, cost respectively £105 and £79 ; the *can-buoys* are from five to eight feet long, and cost from £13 13s. to £20 5s. Smaller buoys are also used in narrow estuaries or rivers. Large buoys are often built on *kneed* frames, resembling the timbers of vessels. The cask-buoy is generally six feet long, and costs £22 15s. All those buoys are formed of strong oaken barrel-staves, well hooped with iron rings, and shielded with soft timber ; and the nozzle-pieces at the small end of the *nun* and *can*-buoys are generally solid quoins of oak or iron, formed with a *raglet* or groove to receive the ends of the staves. Much skill on the part of the cooper is required in heating and moulding the staves to the required form; and great care must be taken that they be of well-seasoned timber. Buoys are not caulked with oakum, but with dry flags, which are closely compressed between the edges of the staves, and swell on being wet ; and they are carefully proved by *steaming* them like barrels, to see if they be quite tight. Sheet iron is sometimes used in making buoys, and they are then sometimes protected with fenders of timber; but they have been found more troublesome for transport, and, for most situations, are considered less convenient than those of timber. An attempt has lately been made, under my direction, to construct buoys of *gutta percha*, stretched on a frame of timber; but I cannot at present speak confidently of the result.

In the beginning of 1845, I suggested the idea of rendering beacons and buoys useful during night, by coating them with some phosphorescent substance, or surmounting them with a globe of strong glass filled with such a preparation, whose combustion is very slow, and emits a dull whitish light and little heat. Some experiments were accordingly made; but no practically useful result has been obtained.

In laying down beacons or buoys, their position is fixed either by the intersection of two lines drawn through two leading objects on the shore, (the magnetic bearings of which are given for the sake of easy reference on the spot, in finding out the marks,) or by means of the angles contained between lines drawn to various objects on the shore, which meet at the beacon or buoy from which they are measured by means of a sextant. In the latter case, the angles are always measured around the whole horizon, thus affording a check by the difference of their sum from 360°. The magnetic bearing of one of those lines is afterwards carefully ascertained by means of the prismatic compass, (if possible from one of the objects on shore, and if not, conversely from the beacon or buoy,) so as to afford the means of translating the whole into magnetic bearings for the use of seamen. The buoys are moored by means of chains and iron sinkers, with a sufficient allowance in the length of the chain to permit them to *ride* easily. * * * * * * * * * *

[Extract from the Journal of the Franklin Institute.]

AN ACCOUNT OF THE CONSTRUCTION OF THE NEW LIGHT-HOUSE AT THE PORT OF HAVANA, CUBA.

By Senor Don Jose Benites, Colonel of Royal Engineers.

[Translated from the Spanish for the Journal of the Franklin Institute]

The improvement of the light at the Moro Castle, which serves as a guide to vessels entering the port of Havana, in such manner as to correspond not only with the importance of its commerce, but also with the perfection to which the construction of this kind of apparatus has now attained, having been determined upon by the royal junta of protection, the department of the marine was naturally consulted with regard to the most eligible situation and the proper altitude of the new light.

That illustrious body was of opinion that a light placed at the same entrance to the port as the former one would be preferable to a light elsewhere, although it might be situated on a more elevated point of the coast, owing to the particular configuration of which it might be seen from further to the windward. With regard to the height, it was judged sufficient to give an additional elevation of twenty-five feet to the old tower, because the light, being thus placed at one hundred and forty-two feet* above the level of the sea, would over-look the point called Del Pajonal, which bears about N. 71° E., and might, therefore, be readily distinguished and recognized by vessels approaching the port from that direction, within not less than eight nor more than fifteen miles of the coast, and affording still greater ad-vantages to those standing in more from the northward. And, more-over, that with the proposed elevation, this light-house would be one of the highest which are known.

In accordance with this report, the junta determined, on the 22d of April, 1840, to proceed with the execution of the work ; and its president, the most excellent Captain General, directed that the corps of engineers should make the proper examination, in order to ascertain the practicability of giving to the old tower the desired ad-ditional elevation. By a report of the commandant, June 2d, 1840, it was declared practicable to add the proposed twenty-five feet to the height of the tower, provided the work were carefully done, and with materials of the best kind.

In this state the project rested until the 16th of October, 1843, when the junta, through its president, asked that the funds necessary for the work should be remitted, inasmuch as it would also be neces-sary, in addition to the projected increase in the height of the tower, to order from Paris a lenticular lantern of the most improved kind, by Fresnel, to be placed therein. The requisite funds having been provided, it was resolved, on the 7th of April, 1844, to proceed with the work, but with the indispensable condition that the light should not be interrupted for a single night. Accordingly, the construction of a small temporary tower was commenced on the 22d of May, for the purpose of sustaining the light while the height of the old tower was being increased by additional mason-work. The great height of the platform upon which the temporary tower was constructed, rendered it unnecessary to give it an elevation of more than seven yards in order to maintain the light at the same height as that at which it was placed in the old tower. Its immediate proximity to this, and the removal of the light to the temporary tower in a single day, rendered

* The Spanish foot here used is equal to 11.1 inches, and the *vara*, or yard, is 33.3 inches.

it unnecessary to give that notice in the public papers which would otherwise have been requisite in order to avoid dangerous consequences to vessels. On the 23d of July, 1844, the temporary lighthouse was first put into use; its cost, including the lantern, amounting to the sum of $1,800. Its strength, and the judicious principles upon which it was constructed, were fully proved in the terrible hurricane of the 5th of October last; during which, though exposed, without the least protection from other buildings, to the force of the wind and the beating of the waves, it remained firm, and without other injury than a total destruction of the glass in the lantern.

The next step was the taking down of the lantern from the old tower, and to proceed with other preparations for increasing its height, when, on the 14th of August, 1844, notice was received by the board of engineers, from his excellency, the Captain General, that, by a resolution of the junta, adopted in virtue of a report from their commissioners, it was determined to suspend the work, and inquire whether it would not be more expedient, provided the funds were sufficient, to construct a tower upon a new plan, which should combine all the beauty, convenience, and facility which would be required for the proper management of an apparatus so complicated as that of Fresnel; advantages which could not be afforded by the old tower on account of its limited dimensions, and its total want of accommodations for the persons entrusted with the care of the light during the night, as well as its narrow and inconvenient stairway. The board of engineers, as was to be expected, reported in favor of the new project, and at the same time submitted a plan and an estimate for the new tower. Both of these were approved by the junta at its session on the 16th of August, and the work was ordered to be carried into effect. On the following day the demolition of the old tower was commenced, and, at the same time, the excavation of a foundation for the new one was begun, in order that the old materials might be used in it.

Situation of the new tower.—Of the rock upon which the Moro castle is built, that point which extends farthest towards the N. W., and on which the old tower stood, is divided from the rest by a cleft or fissure, and is also undermined by a large cavern, washed out by the continual beating of the waves. In order to avoid these defects, which might, in time, endanger the stability of the new edifice, and also because the space afforded for the ground plan was not sufficient, a place was chosen eighty-four feet farther back than the position occupied by the old tower, and on the broadest part of the glacis of the Morillo—a position combining all the advantages of the former one, besides leaving free the extreme point of the Morillo—for the erection of a battery of three Paixhan guns, of large calibre, in such a very advantageous position. Care was also taken to leave sufficient space about the new tower for the free use and management of the cannon which defend the entrance to the port, as well as of those on the opposite side.

Description of the tower.—It is composed of two parts; the first representing a column seventy-nine feet high, twenty-five feet in diameter at the base, and twenty at the top. The cornice of this serves as the floor of a corridor with a circular parapet, enclosed by a grated railing of copper, which surrounds the upper, or second part, upon which the lantern is supported. The first or lower part is constructed of hewn stone, the wall being seven feet thick at the base, vertical in the interior, and sloping on the outside, thirty inches in the whole height. The interior space of eleven feet in diameter serves for a circular staircase, the steps of which are four feet long, seven and a half inches high, nineteen inches broad at the wide end, and six at the other, or immediately at the spindle or central column of three feet in diameter, which extends up through the whole height. The stairway begins inside, at thirty-eight inches in the clear from the door, in order to leave an open landing place ; the steps being covered with slabs of marble one and a half inch in thickness, with a moulding which extends an inch beyond the step. The doorway in the lower part of the tower, which gives entrance to the stairway, is four feet wide and eight feet high, with pilasters at the sides, and its cornice and frontispiece in the form of a circular arch. Over the door is a block of marble containing an inscription in Spanish of the following purport :

In the year 1844 :
Isabella the Second reigning ;
The Junta of Protection under the Presidency of
the Captain General of the Island, Don Leopold O'Donnell ;
This work was executed under the direction of
the Corps of Engineers of the Army.

The stairway is lighted and ventilated by three sets of windows, placed equi-distant from one another throughout the whole height. They are four feet high and two feet wide, and in six of them the lower part of the niche, or recess in the wall which forms the window, is level with the steps, and serves as a landing or resting place. Below the stairway, and at the level of the lower floor of the tower, is a spacious room, enclosed by a wooden railing, which is used as a place of deposit for oil and the more heavy and bulky articles used about the light-house.

At the height of sixty-three feet nine inches the stairway, which has been described, terminates in an apartment called the attendant's room, twelve feet in diameter and twelve and a half in height, covered by an arch two feet in the crown. This chamber, lined with marble, is intended for the two persons who have charge of the light during the night. It is furnished with windows, and the requisite conveniences for containing a supply of lamps, funnels, wicks, oil, &c., &c., having a staircase of mahogany, with a balustrade attached to the wall. The breadth of this stairway is twenty-four inches; the steps are a foot high and eight inches broad in the middle; it leads, through an opening left in the arch to the platform, or upper part of the lower edifice.

Upon this platform is erected a second structure, consisting of a circular wall of hewn stone, two and a half feet thick, eight feet high, and eleven feet seven inches in its interior diameter. The latter dimensions are determined by the base and diameter of the lantern, which, as has been before said, rests upon it. In this second part of the tower there is a door six feet high by twenty-seven inches wide, to admit of egress to the platform, or corridor, above the cornice. In the floor of the apartment within the upper edifice, which is called the lantern room, and which is also lined with marble, there are two apertures; one in the centre, which is circular and twelve inches in diameter, into which is introduced the foot or support of the apparatus; the other is a foot and a half long by two inches wide, for the passage of the cord to which the moving weight is suspended. This weight is raised by means of a pulley placed in the arched ceiling of the attendants' room, and is enclosed in a case thirteen inches square and twelve yards long, formed in the thickness of the wall of the lower part of the tower.

The total height of the tower is, therefore, eighty-seven feet; which, added to the elevation of its base above the level of the sea, sixty-five and a half feet, and to the height of the light within the lantern, five and a half feet, will make the total elevation of the light one hundred and fifty-eight feet, the horizontal tangent of which is thirteen and a half marine miles. This exceeds, by sixteen

15

feet, the height desired in the marine department, and the light is consequently visible about a mile further than the distance required by them.

Description of the apparatus for lighting.—In order to give a clear and connected idea of this, we shall omit a detailed account of the numerous parts of which the various portions of this complicated machinery consist, and shall proceed to describe only such of them as will serve to make the mechanism understood, even by those who are not entirely familiar with the mechanical and optical principles upon which it is constructed.

The new apparatus for lighting consists of four distinct parts, as follow:

1. The mechanical lamp.

2. The system of prismatic lenses and reflectors placed around the light.

3. The machine which gives a uniform rotary movement to the prismatic lenses.

4. The glass lantern which covers the whole without obscuring the light.

The mechanical lamp.—This lamp, placed in the centre of the lenses and reflectors, is supported by a hollow pillar of bronze, seven feet two inches high, having, below, a receptacle for the oil, and a set of pumps for raising it. These pumps are worked by machinery similar to that of a clock, the moving weight being contained within the pillar, and are capable of raising four times the quantity of oil which the burners require; from which it follows not only that the flame is maintained with all possible brilliancy, but also that the superfluous oil which flows over the exterior surface of the tubes and falls back into the cistern, cools the points of the tubes, which might otherwise be fused by the intensity of the heat. In order to increase the watchfulness of those who have the care of the light it is furnished with alarm bells. The escapement of this machinery is connected with one end of a lever, and at the other is suspended a vase with a small hole in the bottom. This vessel is so placed as to receive the superfluous oil from the burners, and, while full, it sustains the counterpoise, but as soon as it begins to vary on account of a deficiency of oil, the lever loosens the movement of the alarm apparatus, which gives notice of the irregularity. A first class light has three lamps, so that if one of them shall be out of order, there are still two fit for use, in good condition, and the light is constantly maintained. The Havana light has four.

The prismatic lenses and reflecting mirrors.—The lenses and reflectors, the object of which is to concentrate the rays of light which proceed from the burners of the lamp, in order to give them a horizontal direction, are placed in a metallic frame conveniently arranged, and which, as well as the lamp, rests upon the capital of the bronze column. One part of this frame, which contains the reflectors, is stationary, while the other, containing the prismatic lenses, revolves around the lamp with a uniform motion. This revolution of the lenses causes eclipses of the light, according to the different positions which they successively occupy around the burner, producing a more intense and lively brilliancy in the direction towards which the focus of each lens happens to be directed, continuing thirty seconds before it is replaced by the next; though, on account of the stationary portion of the apparatus which contains the reflectors, there is always afforded, in all directions, a less brilliant light, but one which is still sufficient to enable the light-house to be kept in sight.

The machinery which gives motion to the prismatic lenses.—The mechanism of this is like that of a clock, and is placed at the side of the bronze pillar which supports the lamp, A pinion-wheel, gearing into another connected with the movable portion of the apparatus, puts it in motion by the action of a weight suspended to a cord passing through the floor of the room which contains the machinery, and descending below into a case made in the wall of the tower.

The lantern.—The glass lantern which encloses the light rests upon the circular wall built above the cornice of the tower, and has the figure of a prism of sixteen sides. It is covered by a cupola of copper, with a chimney in its top to carry off the gases proceeding from the combustion; the whole being surmounted by a weathercock.

By the foregoing description, it will be perceived that this lighthouse, when illuminated, presents constantly a steady light, uniformly alternating, every six seconds, with brilliant flashes, by which it may be distinguished from any other light.

The intensity of the steady light is equal to that given by five hundred and fifty burners like those of the ordinary Carcel lamp, which consume three and a quarter ounces of oil per hour, and it may be easily seen at a distance of six or seven marine leagues. The brightness of the flashes is nearly four times as great as that of the steady light, and equal to that given by two thousand of the abovementioned Carcel burners.

Data for determining the proper thickness for the wall of the tower.—
The most interesting and nice question which first presents itself, in
considering the construction of a very high tower, is to determine
the thickness which its walls ought to have in order to insure the
necessary stability, and to resist successfully the forces which have
a tendency to destroy it. This is a general principle in all kinds of
edifices; but in high towers, the base of which bears a small proportion to their height, it is necessary to consider another element of
destruction more powerful and active than the force of gravitation,
namely, the violence of the wind in those great hurricanes which are
so frequent on seacoasts, particularly among the Antilles.

In the calculation of this force, as in all physico-mathematical
questions, the data are, for the most part, derived from experiments,
of the accuracy of which we cannot be precisely certain, but which
have been found sufficiently worthy of confidence when applied to
similar structures which have for many years resisted every kind of
force; and in this case there is nothing more logical or natural, than
to take the result of these calculations as the established data of
comparison with reference to our own case.

The data upon which we should reckon in calculations of this kind,
founded upon numerous experiments, are as follows: First, the pressure upon a superficial metre, when the wind has a velocity of fifty
metres per second, which is that of the greatest hurricanes, is equal
to the weight of two hundred and seventy-five kilogrammes, or about
three hundred and eighty-six pounds to the square yard; second, the
action of the wind upon a vertical cylinder is reduced to two-thirds
of the force which it exerts upon a central vertical section of the
same cylinder.

To apply these rules to the stability of a tower composed of a single stone in the form of a right prism with a square base, we should
multiply the superficial area of one of its sides by three hundred
and eighty-six pounds; and this product being multiplied by half the
height of the prism, will give the momentum of the wind's force.
In order to determine the momentum of resistance, the total weight
of the tower should be multiplied by half the width of its base. Dividing this product by the first obtained momentum of the wind, we
have the expression of the *absolute stability* of the edifice with regard
to the action of hurricanes.

Towers are composed of numerous courses of cut stone, the adhesion of which affects the stability in proportion to the decrease in

the size of the blocks of stone used. If we suppose this adhesion to be nothing in a wall composed of courses of stretchers and ties, the fracture caused by the force of the wind will not be horizontal; but (as in retaining walls) according to an oblique section, the inclination of which is determined by the maximum of relation between the force of the wind and the momentum of weight of the solid above the section. But it should be observed that in most high towers, the adhesion of the cement and the arrangement of the materials is such that the section of rupture is less inclined than in ordinary containing walls, and if, in addition to this, we also consider that the results which we seek are of comparison between our tower and those whose stability has been already tested by having for a long time resisted, without injury, every kind of force, the supposition that the section of fracture would be horizontal should have no sensible influence.

The determination of the section of most easy fracture offers a problem of minimums, the direct solution of which, in general, requires very tedious calculations; but fortunately the position of that section is apparent, in most cases, at the first glance of the eye, or may readily be determined by a few trials.

In case the tower should have the form of a truncated cone, like that of which we are treating, if it should ever yield to the force of the wind, it would be twisted off above the offset of its base, in the direction opposite to that from which it receives the force, and its being broken above the offset would be because the arm of the lever of resistance, instead of being equal to the radius of the section of rupture, is somewhat less.

These principles being established, it is known that, in order to give great stability to a circular wall in ordinary structures, it is sufficient to make the thickness a twelfth part of the height; consequently, having to raise our tower to a height of eighty feet, a thickness of seven feet was enough for the wall; but as the tower is not an insulated wall, but has within it, a circular stairway which binds and fastens together all its parts, and, moreover, as the consideration of a convenient situation for the illuminating apparatus required that the upper part of the tower should be only twenty feet in diameter, we were able to start with seven feet at the base, and to diminish gradually to four and a half at the top, with the advantage of an exterior slope of two and a half feet in the whole height, which also adds to its strength.

Having established these data, it now remains to examine whether, with these dimensions, the tower possesses as much, or greater, strength than others similar to it, which have been built many years, and have resisted, without the least damage, the force of the greatest storms, and I say it ought to be stronger: 1st, Because the terms of comparison which we seek are derived from the towers built in Europe, where hurricanes are neither so frequent nor of such long duration as in the Antilles; and, 2d, Because the tower which we have constructed, from its peculiar situation, has to sustain not only the force of the wind but the shock of the waves, although this is somewhat diminished by the tower being placed twenty-two yards above the level of the sea.

For comparison, we will take the signal tower erected in 1715, at the port of L'Orient, the form of which is similar, on the whole, to that constructed at the Moro, and which has withstood until the present time, without injury, all the force of the storms.

Stability of the tower at L'Orient.—The result of a calculation with regard to this tower shows that the relation of the resistance to the pressure is 7.4; that is to say, the resistance is 7.4 times greater than the pressure. Let us now see what result will be obtained by an analogous calculation with regard to the tower at the Moro.

Calculation of the stability of the tower at the Moro.—The contents of the truncated cone which constitutes the lower part, is 31,174 cubic feet. The interior cylinder, which contains the staircase, is 11 feet in diameter and $34\frac{1}{2}$ in circumference; the area of its base, 95 square feet; and its contents $95 \times 78 = 7,410$ cubic feet. Taking this from the total solid contents, there remain 23,764 cubic feet for the solid contents of the circular wall of the lower part of the tower. The cylindrical column in the centre of the circular stairway contains 450 cubic feet, and the stairway itself 306; so that the total solidity of the masonry in the lower body of the tower is 24,520 cubic feet, or 908 cubic yards. Adding to this three cubic yards for the projection of the cornice, and subtracting thirty-six yards for the apertures left for doors and windows, there remain 875 cubic yards of masonry. As the attendant's room, in the lower section of the tower, is a foot larger in diameter than the interior space occupied by the stairway, it follows that the amount of masonry, which on this account has been rather over-estimated, will be made up by the arched ceiling of this room, which has not been taken into the account.

The solid contents of the circular wall which forms the upper section of the tower, upon which the lantern is placed, is twenty-seven cubic yards, deducting the opening for the door; so that the whole amount of solid masonry is 902 cubic yards.

A cubic yard of the masonry used in this tower weighs eighty-five arrobas, (twenty-five pounds each,) and the arm of the lever of resistance being four yards, (a little less than the radius of the base,) it follows that the power of the resistance will be $902 \times 4 \times 85 = 306,680$.

A central perpendicular section of the lower portion of the tower contains 195 square yards; of the upper portion, 13; and of the lantern, 26, presenting a total sectional superfice of 234 square yards. The centre of gravity of these three superfices is found at thirteen yards from the base of the tower; consequently the momentum of pressure from the wind is $\frac{2}{3} 386 \times 234 \times 13 = 31,504$ arrobas, and the ratio of the resistance to the pressure $= \dfrac{306,680}{31,504} = 9.7$.

By this result it will be seen that the tower at the Moro has a stability equal to nearly one and a half times that of the tower at L'Orient, which was taken for comparison, and consequently that its dimensions are sufficient, although indispensable for the reasons already mentioned, to warrant the undertaking of its construction with entire confidence.

Details of the construction. Foundation.—As the tower is twenty-five feet in diameter at its base, a circle 28½ feet in diameter was traced around the centre of the area to be occupied by it so that the excavation for the foundation being larger than the body of the tower, a berm or offset of twenty-one inches would be left by the extension of the foundation wall beyond that of the tower itself. The excavation was carried through the surface soil down to the rock which underlies the glacis of the Morillo, which was reached and laid bare at an average depth of fifteen feet.

The rock presented an irregular surface, in consequence of which, together with its extreme hardness, the operation of cutting it down to a level would have been very expensive. It was therefore deemed preferable to construct the foundation by steps, (*en escalones*,) using blocks of stone of the best kind, disposed in circular horizontal layers, (*formando anillos en capas horizontales*,) properly placed, and the joints filled with a mixture of lime and sand; taking care, as in all masonry of hewn stone, that the joints of each layer should be covered by the course immediately above it.

The foundation is nine and a half yards in diameter by five in depth, and consequently contains three hundred and fifty-five cubic yards of masonry. The cost of its excavation and refilling was $1,554, being $4.375 per cubic yard. In the calculation of the original estimate it was assumed that the foundation would be only three yards in depth; consequently containing two hundred and thirteen cubic yards, of which the cost, estimated at $7.50 per yard for the masonry, including the excavation, amounted to $1,597.50, being an excess of $43.50.

In calculating this estimate of $7.50 per cubic yard of masonry, it was proposed to use all the materials of the old tower: but as this would not be sufficient, it was supposed that the stone still to be provided would have to be purchased and transported from the opposite shore. The great reduction which was made in the cost per cubic yard of this work, from that originally estimated, was owing to the following causes : *First.* Instead of buying from individuals the stone which were required, those were procured which remained unused from the erection of the fortifications at that place, and which has been taken from quarries opened for that purpose ; on which account they were obtained at a much lower price, while, at the same time, the expensive transportation in vessels was avoided. *Second.* Because it was not necessary to purchase sand for making the cement, having used that which was found in the fosses of the fortifications.

The digging of the foundation was commenced on the 17th of August, 1844, and by the 24th of September it was completed and leveled off to nine inches below the glacis of the castle ; being then in order for receiving the first layer of stone for the tower. During the same time the old tower was completely taken down, its materials, as has been already said, being used for the new foundation.

The lower part of the tower.—Upon the level top of the basement, or foundation, the first layer of stone forms a solid floor over the whole twenty-five feet in diameter ; its outer boarder being vertical and forming a projection or socle of nine inches around the tower, as well as a step, of the same height, for entrance into it. The building is then continued to the requisite elevation by successive courses of eighteen inches in height. In order to give an exterior slope to the wall, and consequently to diminish its thickness in due proportion, the width of the blocks in the interior courses was gradually contracted, while those on the exterior were of the same breadth, so that the vertical joints might recede gradually towards the interior, and the masonry be well united together; thus avoiding the defect which would result from an equal diminution in the breadth of the blocks used in both the inner and outer faces of the wall, in which case, the vertical joints of each course would continue to be one above another, thus forming two distinct walls, without further connexion than that of the extending ties.

The circular stairway is composed of entire blocks of stone, which are inserted six inches into the wall, each of them forming two steps, together with the corresponding part of the central column, or nave, of the staircase. It was carried up at the same time with the wall of the tower, in order to insure the proper connexion and binding together of the work, and also served for use during the construction of the building; the order of proceeding being first to set the block of stone which formed two steps, and then to lay the corresponding course on the wall, and so on successively. The nave, or central part of the stairway, is cylindrical throughout its whole height, except at the top, where it increases in diameter, with a view of gaining a certain space on the floor of the attendants' room. In order to give this proper form to the central column, its diameter was gradually enlarged, beginning at the seventeenth step from the top, and so continuing to the last; so that the stairway is only twenty-seven inches wide in the last six feet of its flight.

The upper part of the tower.—This is a simple circular wall of cut stone, composed of blocks in alternate order, and presenting nothing worthy of special notice.

Arch of the attendants' room.—This arch is in the form of a simple dome, and presents nothing remarkable in its construction. The thickness of the wall which supports it is that which is strictly necessary for the equilibrium, since the excess in the weight of its masonry, from the spring of the arch to the key, and that of the upper part of the tower, which also contributes to its stability, are more than sufficient to compensate for the weight of the apparatus for illumination, which rests upon the summit of the arch. But notwithstanding this, as the three openings in it somewhat diminish its strength, the precaution was taken, in order to avoid the least displacement, to fasten all the arch stones of the second and third courses, together with those of the first, with iron clamps, one and a half inch thick by three inches wide, let into the arch stones and fastened to each of them by a bolt, which passes through two-thirds of their thickness, well secured with lead.

Materials.—The stone used in the construction of the tower was taken from the quarries known here by the name of *playa de chivos* immediately at the point where the tower stands, and preferable to the stone found at any other place in the vicinity, because, although not of the hardest quality, they are still sufficiently so, and of a more equal and homogeneous texture than the others. Great care was also taken to use blocks of the most durable kind in the exterior surface of the wall, and for the steps of the stairway.

The lime and sand employed were both of excellent quality; the former made of a very hard and clean stone, and the latter containing no saline matter. The mortar was made with great care, composed of two-fifths of lime and three-fifths of sand, mixed with fresh water and well stirred, so as perfectly to incorporate the ingredients.

The doors, windows, interior balustrades, and hand rails of the staircase are of mahogany.

After the completion of the edifice, in order to preserve it from the effects of moisture and of nitrous salts, inasmuch as the stone, as already stated, is not of the hardest kind, it was covered, both on the interior and exterior surface, with a very thin and well laid coat of painted stucco.

The cost of the structure, when completed, was very nearly that calculated in the original estimate.

CHAMBER OF COMMERCE, NEW YORK.

The Committee appointed to consider the condition of the light-houses of the United States, as developed in the report of the Light-house Board recently communicated to Congress, ask leave to report:

That the subject has attracted occasional attention for some years, although your committee believe it never received a thorough investigation until the Light-house Board was appointed by the Secretary of the Treasury in May, 1851, under authority given by Congress. In January, 1838, the Senate called on the Treasury Department for a copy of a communication made by Messrs. E. & G. W. Blunt, relative to the light-houses of the United States, and also for a copy of the report in reply to that communication, made by Mr. Pleasanton, the Fifth Auditor of the Treasury, he being in charge of the whole light-house establishment. In the communica-

tion of Messrs. Blunt, they expressed the opinion that our establishment was badly managed, and was greatly inferior to the similar
establishments of France and Great Britain. In his reply, Mr.
Pleasanton undertook to show that this opinion was not well founded,
and that the charges accompanying it were frivolous, unimportant,
or groundless. Messrs. Blunt rejoined, with a view to prove that
their statements were correct. In their first communication, Messrs.
Blunt took occasion to state that the French light-houses were superior to the English, the former being provided with lenses instead of
parabolic reflectors.

In May, 1838, the Committee on Commerce of the Senate reported
upon the expediency of introducing the dioptric apparatus for illuminating light-houses into the United States, and upon the expediency of improving the organization of our light-house system. In
that report it is stated that the Navy Board was required at the previous session of Congress to examine the sites of a large number of
new light-houses authorized at that session, and inquire into the facts
concerning them. The act of Congress provided that in cases where
objections were found to exist the erection of the buildings should be
suspended. Under this provision the Navy Board arrested the erection of no less than thirty-one of the houses proposed.

The committee further stated that while our light-house system
(if system it could be called) had, probably, most of the time, been
conducted with reasonable care and satisfaction to the public in most
respects, both the French and English had, by scientific research
and improvement, perfected theirs to a considerably higher degree.
The French took the lead, under Fresnel, who brought the lenticular
apparatus into use. The committee were satisfied, with the evidence before them, that this apparatus was the best known, but to
test the matter practically, they recommended that two sets of the
first class and one of the second should be imported and put up at
proper points to try their merits by actual experiment.

Captain W. H. Swift, of the United States Topographical Engineers, stated, in 1842, that he had never seen upon our coast a light
which approached, even in a remote degree, the brilliancy exhibited
by certain English lights which he referred to. Fresnel's apparatus
he considered superior to the English, and both immeasurably superior to our own.

In 1845, two officers of the navy, viz: Lieutenants Jenkins and Bache, were sent to Europe by the then Secretary of the Treasury, to obtain information concerning the light-house establishments there. In the report of these officers, on their return, they denounced our lights as inferior to all they had seen abroad; recommended strongly the adoption of the French apparatus, and the appointment of a board of officers to devise some new plan of light-house management.

A committee of the Franklin Institute, of Philadelphia, in May, 1849, stated that they were strongly impressed with the great excellence of the Fresnel system, a system which had then been established, as they said, in nearly three hundred places in Europe, where it was gradually superseding the old lights, whilst in no instance had it been abandoned after it had been once established.

The recent Light-house Board consisted of Commodore Shubrick and Commander Du Pont, of the Navy, General Totten, of the Engineers, Colonel Kearney, of the Topographical Engineers, and Professor Bache, Superintendent of the Coast Survey, as members, with Lieutenant Jenkins, of the Navy, as Secretary. Their report sets forth, among other things, that in 1825, the French government adopted definitively what is called the French system of illumination, and that in 1834, a new impulse was given in England and Scotland to light-house improvement by letters from Sir David Brewster, and by the action of a committee of the House of Commons. Although, say the Board, the lens met with much favor in England, and has been gradually getting into use, until nearly one-half the seacoast lights have been changed since 1837, still Scotland has introduced a larger number in proportion to extent of coast. Following out these improvements, another committee on light-houses was raised by the House of Commons in 1845, and the result, among other alleged benefits, has been the introduction of a large number of lens apparatus, not only into Great Britain but into many of the Colonies.

The Board, however, have been unable to discover that any steps have been taken in this country to keep pace with the light-house improvements of France and Great Britain, except the placing a lens apparatus at the Highlands of Navesink, an improved reflector apparatus in the Boston light-house, a lens of the second class at Sankaty Head, Nantucket, and the lights authorized by law to be constructed, under the direction of the Topographical Bureau, at

Brandywine Shoal, Carysfort Reef, and Sand Key. The law required that the lenticular apparatus placed at Navesink, and the improved reflector apparatus at Boston should be tested by full and satisfactory experiment, as to their merits, in comparison with the other apparatus in use. But the Board have never heard the result, nor that any such experiment was ever made.

The Board, after examining the subject committed to them with the patience and zeal its importance demanded, arrived at a great number of conclusions set forth with remarkable particularity in their report. They condemn most emphatically our whole system. They distinctly declare, that in their judgment the light-houses, light-vessels, beacons, and buoys, and their accessories, in the United States, are not as efficient as they ought to be, and not so much so as those of France and Great Britain. That our light-house establishment does not compare favorably in point of economy with theirs. That the towers and buildings are not constructed, in general, of the best materials, nor with proper accommodations. That the lanterns, as a general rule, are of improper dimensions, constructed of ill-adapted materials, without scientific skill, and are in many instances not suited to the use designed. That there is no systematic plan of construction, illumination, and superintendence. That the illuminating apparatus is of a description now nearly obsolete where the best of that kind was employed before the introduction of the French lenses. That the seacoast reflector lights are in general too low, and are deficient in power and range. That the attendance is deficient. That the lights are not properly classified, nor well distinguished from each other. That there is not in useful effect a single first class light on the coast. That the lights at Navesink and Sankaty Head, Nantucket, are the best. That the Fresnel lens is greatly superior to any other mode of light-house illumination. That there is no proper system of distributing supplies to light-houses, and no system in the management of the light-house establishment. That the light-keepers in many instances are not competent. That the mode of repairs is not efficient or reliable. That the floating lights are comparatively useless for want of efficient lamps and reflectors. That the light-vessels are not well adapted to the service, and are not properly distinguished either by day or by night. That the buoys are defective and the moorings insufficient. And that there is no proper system of beaconage or buoyage.

These are but a few of the conclusions of the Board. Yet these are quite enough, if well founded, to show that the condition of the whole light-house establishment is such as to demand thorough renovation and reform. It is not to be endured that the lives of our seamen, and of the passengers often embarked with them, should ever be put in peril for want of the best protection in approaching our ports, or proceeding along our coast, that it is in the power of skill, science and systematic management to give. Upon no consideration whatever ought the United States to be second to any country under heaven in affording all possible security to the vast interests of commerce and navigation, upon which so much of national prosperity depends.

For the correctness of the positions taken by the Light-house Board, and the sufficiency of the evidence on which they are based, it seems to your committee that the high character, elevated standing, and eminent attainments of the members of the Board furnish an ample guarantee. If corroborative testimony were needed, more than enough, in the opinion of your committee, may be found in the various documents referred to in this report. Still it is due to Mr. Pleasanton to state, that in his reply to the report of the Light-house Board, he considers he has shown that our lights are satisfactory to the captains of ships and pilots generally; that our light-house establishment is managed more economically than that of Great Britain; that our light-ships are superior to the British; and that the French lenses are more expensive, without showing a better light than the reflectors. It thus appears that the parties stand in direct opposition to each other. Both cannot be right. Your committee, after duly considering the question, cannot avoid the conclusion that Mr. Pleasanton is far more likely to be in error than the Light-house Board.

By way of remedy for the evils they specify, the Board recommend among other things the absolute adoption of the Fresnel apparatus for all new lights and all lights requiring renovation, modified in special cases by Stevenson's apparatus. They also recommend the organization of a permanent Light-house Board for the construction, repair, management, and superintendence of the light-houses, light-vessels, beacons, and buoys, of the United States. They propose that this Board be composed of the Secretary of the Treasury, as President, two officers of the Navy, of high rank, one officer of Engineers, one officer of Topographical Engineers, and two civilians of high scientific attainments, as members, with an officer of the Navy and an officer of Engineers as secretaries; the officers, civil, military, and naval, to serve without additional salary.

Your committee cannot doubt that the organization of a Light-house Board of this description would secure, beyond question, the selection of proper sites for light-houses along our shores, with due regard to the wants of commerce. They are satisfied it would also render quite certain the proper construction of the light towers and buildings, as well as the use of the best illuminating apparatus. They are convinced, too, it would ensure far greater efficiency, economy, and systematic management throughout the whole light-house establishment, than is practicable under the superintendence of an auditor of the Treasury Department. A board of skilful, scientific men, combining the variety of qualification required, is obviously preferable for these purposes, to a disbursing officer, not supposed to be very highly endued with engineering, nautical, or optical knowledge. The fact stated by the Committee on Commerce of the Senate, in 1838, that the Navy Board stopped the erection of thirty-one new light-houses in one year, for sufficient reasons, indicates very clearly the necessity of a proper light-house organization at all times, even if no stress be laid upon the conclusions of the recent able Light-house Board. This necessity is certainly not lessened by the further facts that the present superintendent has omitted to test the merits of the Fresnel apparatus, as virtually directed by the act of Congress of 1838, and that he still strenuously opposes the more frequent selection of that apparatus for the use of our light-houses, in the face of a weight of testimony in its favor, which to your committee appears irresistible. This course plainly discovers a steadfast determination on his part against reform and improvement, far removed, as your committee think, from the true spirit which should actuate the head of so important an establishment. In any view your committee are able to take of the subject, the proposed change seems highly expedient. They therefore recommend the adoption of the accompanying resolution.

All which is respectfully submitted.

GEO. CURTIS,
LEOPOLD BIERWIRTH, } *Committee.*
JAMES W. PHILLIPS,

Resolved, That the Chamber of Commerce of the City of New York cordially approves of the recommendation of the Light-house Board, contained in their report of 30th of January last, that a permanent board be organized, to be charged by law with the entire management of the light-house establishment of the United States; and that the Chamber sincerely hopes that Congress will carry that recommendation into effect.

At a meeting of the Chamber of Commerce at New York, held on the 4th May, 1852, the preceding report and resolution were approved, and ordered to be printed and sent to the Senate and House of Representatives as the sense of the Chamber.

MOSES H. GRINNELL, President.

M. MAURY, *Secretary.*

ROYAL INSTITUTION OF GREAT BRITAIN.

WEEKLY EVENING MEETING.—FRIDAY MARCH 9, 1860.

The Lord Wensleydale, Vice President, in the Chair.—Professor Faraday, D. C. L. F. R. S.

On Light-house Illumination—The Electric Light.

The use of light to guide the mariner as he approaches land, or passes through intricate channels, has, with the advance of society, and its ever increasing interests, caused such a necessity for means more and more perfect, as to tax to the utmost the powers both of the philosopher and the practical man, in the development of the principles concerned, and their efficient application. Formerly the means were simple enough; and if the light of a lanthorn or torch was not sufficient to point out a position, a fire had to be made in their place. As the system became developed, it soon appeared that power could be obtained, not merely by increasing the light but by directing the issuing rays: and this was in many cases a more powerful and useful means than enlarging the combustion—leading to the diminution of the volume of the former with, at the same time, an increase in its intensity. Direction was obtained, either by the use of lenses dependent altogether upon refraction, or of reflectors dependent upon metallic reflexion; and some ancient specimens of both were shown.

In modern times the principle of total reflexion has also been employed, which involves the use of glass, and depends both upon refraction and reflexion. In all these appliances much light is lost: if metal be used for reflexion, a certain proportion is absorbed by the face of the metal; if glass be used for refraction, light is lost at all the surfaces where the ray passes between the air and the glass; and also in some degree by absorption in the body of the glass itself. There is, of course, no power of actually increasing the whole amount of light, by any optical arrangement associated with it.

The light which issues forth into space must have a certain amount of divergence. The divergence in the vertical direction must be enough to cover the sea from the horizon, to within a certain moderate distance from the shore, so that all ships within that distance may have a view of their luminous guide. If it have less, it may escape observation where it ought to be seen; if it have more, light is thrown away which ought to be directed within the useful degree of divergence; or if the horizontal divergence be considered, it may be necessary so to construct the optical apparatus, that the light within an angle of 60° or 45° shall be compressed into a beam diverging only 15°, that it may give in the distance a bright flash having a certain duration instead of a continuous light—or into one diverging only 5° or 6°, which, though of far shorter duration, has greatly increased intensity and penetrating power in hazy weather. The amount of divergence depends in a large degree upon the bulk of the source of light, and cannot be made less than a certain amount, with the flame of a given size. If the flame of an Argand lamp ⅞th of an inch wide, and 1½ inch high, be placed in the focus of an ordinary Trinity House parabolic reflector, it will supply a beam having about 15° divergence; if we wish to increase the effect of brightness, we cannot properly do it by enlarging the lamp flame; for though lamps are made for the dioptric arrangement of Fresnel, which have as many as four wicks, flames 3½ inches wide, and burn like intense furnaces, yet if one be put into the lamp place of the reflector referred to, its effect would chiefly be to give a beam of wider divergence; and if to correct this, the reflector were made with a greater focal distance, then it must be altogether of a much larger size. The same general result occurs with the dioptric apparatus; and here, where the four-wicked lamps are used, they are placed at times nearly 40 inches distant from

16

the lens, occasioning the necessity of a very large, though very fine, glass apparatus.

On the other hand, if the light could be compressed, the necessity for such large apparatus would cease, and it might be reduced from the size of a room to the size of a hat; and here it is that we seek in the electric spark, and such like concentrated sources of light, for aid in illumination. It is very true, that by adding lamp to lamp, each with its reflector, upon one face or direction, power can be gained; and in some of the revolving lights, ten lamps and reflectors unite to give the required flash. But then not more than three of these faces can be placed in the whole circle; and if a fixed light be required in all directions round the light-house nothing better has been yet established than the four-wicked Fresnel lamp in the centre of its dioptric and catadioptric apparatus. Now the electric light can be raised up easily to an equality with the oil lamp, and if then substituted for the latter, will give all the effect of the latter; or by expenditure of money it can be raised to a five or tenfold power, or more, and will then give five or tenfold effect. This can be done, not merely without increase of the volume of the light, but whilst the light shall have a volume scarcely the 2000th part of that of the oil flame. Hence, the extraordinary assistance we may expect to obtain of diminishing the size of the optical apparatus and perfecting that part of the apparatus.

Many compressed intense lights have been submitted to the Trinity House; and that corporation has shown its great desire to advance all such objects and improve the lighting of the coast, by spending, upon various occasions, much money and much time for this end. It is manifest that the use of a light-house must be never failing, its service ever sure; and that the latter cannot be interfered with by the introduction of any plan, or proposition, or apparatus, which has not been developed to the fullest possible extent, as to the amount of light produced—the expense of such light, the wear and tear of the apparatus employed, the steadiness of the light for 16 hours, its liability to extinction, the amount of necessary night care, the number of attendants, the nature of probable accidents, its fitness for secluded places, and other contingent circumstances, which can as well be ascertained out of a light-house as in it. The electric spark which has been placed in the South Foreland High Light, by Professor Holmes, to do duty for the six winter months, had to go through all this preparatory education before it could be allowed this practical trial. It is not obtained from frictional electricity, or from voltaic electricity, but from magnetic action. The first spark (and even magnetic electricity as a whole) was obtained 28 years ago. (Faraday, *Philosophical Transactions*, 1832, p. 32.) If an iron core be surrounded by wire, and then moved in the right direction near the poles of a magnet, a current of electricity passes, or tends to pass, through it. Many powerful magnets are therefore arranged on a wheel, that they may be associated very near to another wheel, on which are fixed many helices with their cores, like that described. Again, a third wheel consists of magnets arranged like the first; next to this is another wheel of the helices, and next to this again a fifth wheel, carrying magnets. All the magnet wheels are fixed to one axle, and all the helix wheels are held immovable in their place. The wires of the helices are conjoined and connected with a commutator, which, as the magnet-wheels are moved round, gathers the various electric currents produced in the helices, and sends them up through two insulated wires in one common stream of electricity into the light-house lanthorn. So it will be seen that nothing more is required to produce electricity than to revolve the magnet-wheels. There are two magneto-electric machines at the South Foreland, each being put in motion by a two-horse power steam-engine; and, excepting wear and tear, the whole consumption of material to produce the light is the coke and water required to raise steam for the engines, and carbon points for the lamp in the lanthorn.

The lamp is a delicate arrangement of machinery, holding the two carbons between which the electric light exists, and regulating their adjustment; so that whilst they gradually consume away, the place of the light shall not be altered. The electric wires end in the two bars of a small railway, and upon these the lamp stands. When the carbons of a lamp are nearly gone, that lamp is lifted off and another instantly pushed into its place. The machines and lamp have done their duty during the past six months in a real and practical manner. The light has never gone out, through any deficiency or cause in the engine and machine house; and when it has become extinguished in the lanthorn, a single touch of the keeper's hand has set it shining as bright as ever. The light shown up and down the channel, and across into France, with a power far surpassing that of any other fixed light within sight, or anywhere existent. The experiment has been a good one. There is still the matter of expense and some other circumstances to be considered; but it is the hope and desire of the Trinity House, and all interested in the subject, that it should ultimately justify its full adoption.

[M. F.]

THE LIGHT DUES LEVIED ON THE SHIPPING OF THE UNITED STATES IN GREAT BRITAIN.

CORRESPONDENCE RESPECTING THE LIGHT DUES LEVIED ON AMERICAN SHIPPING IN THE UNITED KINGDOM.

No. 1.

Mr. Lawrence to Viscount Palmerston.—(Received January 2, 1851.)

UNITED STATES LEGATION,
138 PICCADILLY, *December* 31, 1850.

The undersigned, Envoy Extraordinary and Minister Plenipotentiary of the United States of America, has the honor, under instructions from his government, to invite the attention of Viscount Palmerston, her Majesty's Principal Secretary of State for Foreign Affairs, to the subject of the light dues exacted of the mercantile marine of the United States entering the ports of the United Kingdom.

It is nearly two years since her Majesty's government determined to respond to the invitation of the United States, made to the world many years ago, and recommend Parliament to repeal the prohibitory navigation laws of the kingdom; and the commerce of the two nations has been conducted for now just one year on the reciprocal basis established in accordance with such recommendation. The United States ask Great Britain to carry out this principle; to establish reciprocity in fact as well as in name, to do justice to the commerce of the United States.

The light-houses, floating-lights, buoys, and beacons on the whole sea and lake-coast, and rivers of the United States, were constructed and are maintained by the federal government, an annual appropriation being made by Congress for these objects. No light dues of any kind are levied, either on vessels of the United States or on ships of foreign nations. In the year 1792 there were but 10 lights in the Union. In the year 1848 there were 270 light-houses, 30 floating lights, and 1,000 buoys, besides numerous fixed beacons. There are probably at this time, including those under construction on the Pacific coast, more than 300 light-houses, with a proportionate number of floating lights, buoys, &c., all of which are given to the use of the world by·the United States without tax or charge.

The commerce of the United States is not met with a corresponding liberality in the ports of the United Kingdom. The reciprocity intended to be established by the navigation law of last year, and so much to be wished for between the two greatest commercial nations of the globe, will not be realized if American tonnage continues subjected to onerous light dues in Great Britain, while British vessels enjoy without pay the lights, &c., upon the coast of the United States.

The undersigned is not unaware that the system of light dues in this country is somewhat complicated, but he believes that as reforms have been made in many other laws and customs adopted in former ages under a different state of society, so changes can be made in these, adapting them more to the present condition of the world and to the liberal policy of other nations.

In illustration of the onerous and unequal operation of the present system of lights in Great Britain upon American commerce, the undersigned has the honor to ask Viscount Palmerston's attention to a few examples:

The American mail-steamers entering at Liverpool pay for light dues the sum of 62*l.* for each voyage. If the British steamers were subjected to the same charge in American ports, it would amount annually to the great sum of 3,224*l.* Two steamers, the "Washington" and "Hermann," returning from New York to Bremen, and touching on the way at Southampton, paid last year light dues to the amount of 800*l.*, including the charge for the Heligoland light, which the undersigned is informed they have never seen. The steamer "Franklin," running between New York and Havre, and touching at Cowes, but without anchoring, merely to land its mails, has been subjected to light dues, which have been paid by order of the Trinity House, but under protest of the consignees. One commercial house in New York, running sixteen ships between that port, Liverpool, and London, paid last year for such dues, 2,498*l.* 3*s.* 6*d.* Another American shipping house paid for lights in 1849 the sum of 2,252*l.*

The undersigned will not multiply individual examples of the amount of this tax. The number of American ships that entered the ports of the United Kingdom for the nine months ending October 1, 1850, was six hundred and two (602,) with a tonnage of four hundred and seventy-three thousand nine hundred (473,900.) If one-third be added for the last quarter of the year, the total will be seven hundred and fifty-two (752) ships, and about six hundred thousand tons, being an average of nearly eight hundred (800) tons to each ship. Assuming this calculation to be substantially correct, it appears from the Trade List, that three thousand nine hundred and nine (3,909) American vessels have entered the ports of the United Kingdom in the last four years, with a tonnage of two millions four hundred and twenty-two thousand four hundred and ninety-two (2,422,492.) During the last ten years, the number of American vessels entered in the United Kingdom has been seventy-eight hundred and seventy-two (7,872,) with an aggregate of four million six hundred and eighty-one thousand nine hundred and twenty-five (4,681,925) tons.

The light dues are not the same in all the ports of the United Kingdom. The rate at Liverpool is ten pence halfpenny (10½d.) or eleven pence (11d.,) whereas the charge in London is thirteen pence (13d.) per ton. If the average is taken at one shilling (1s.) per ton, a ship of eight hundred (800) tons would pay forty pounds (40l.,) and seven hundred and fifty-two (752) ships, with six hundred thousand (600,000) tons register, would pay about thirty thousand pounds (30,000l.,) or one hundred and forty-six thousand dollars ($146,000) annually, and it is not probable that the amount of tonnage will decrease in coming years. Thirty-nine hundred and nine (3,909) American ships must have paid with a tonnage of two millions four hundred and twenty-two thousand four hundred and ninety-two (2,422,492,) the sum of one hundred and twenty-one thousand pounds (121,000l.,) or nearly six hundred thousand dollars ($600,000.) And within ten years last past, the shipping of the United States has contributed, upon seven thousand eight hundred and seventy-two (7,872) vessels, the aggregate tonnage of which was four million six hundred and eighty-one thousand nine hundred and twenty-five (4,681,925,) the immense sum of two hundred and thirty-four thousand pounds (234,000l.,) or over one million one hundred thousand dollars ($1,100,000,) for the support of the light-house system of the United Kingdom.

During the last year there appears to have been levied upon the shipping of the world for the light dues in the United Kingdom, between four and five hundred thousand pounds. Of this, one-fourteenth part was paid by citizens of the United States; while British subjects, with a fleet doubtless equally large in the ports of the United States, have not been taxed at all for the maintenance of lights. The government of the United States, in view of its liberality, is justified in asking her Majesty's government to do away with this great inequality, and remove this restriction on commerce.

The undersigned believes that no other nation levies so heavy a tax upon ships in the form of light dues as this. There are in the United States many more light-houses, &c., than in the United Kingdom; yet the annual appropriation for the construction of new, the repair of old, light-houses, and the maintenance of both, does not equal one-fifth of the annual amount raised for this purpose in the United Kingdom by the tax on the shipping coming into its ports.

In conclusion, the undersigned, on behalf of his government, expresses the wish that her Majesty's government may take this grave subject into consideration; that it may speedily set free American shipping from so unequal and so onerous a restriction ; that it may complete its great measures for commercial freedom; and may thus increase still more the intercommunication which is already producing so happy results.

The undersigned, &c.   (Signed)

ABBOTT LAWRENCE.

No. 2.

Viscount Palmerston to Mr. Lawrence.

FOREIGN OFFICE, *February* 6, 1851.

The undersigned, &c., has the honor to inform Mr. Lawrence, &c., that her Majesty's government has had under its consideration the note which Mr. Lawrence addressed to the undersigned on the 31st of December, complaining that heavy light dues are levied on American shipping in the ports of the United Kingdom, whilst no dues of the same kind are charged upon British ships frequenting the ports of the United States ; and Mr. Lawrence says that he is instructed by his government to request that measures may be taken in order that American vessels may enjoy in British ports, in regard to exemption both from light dues and from other similar charges, the same advantages which British vessels enjoy in the ports of the United States.

In reply, the undersigned has the honor to state, that the difference between the treatment of British vessels in American ports, and that of American vessels in British ports in regard to light dues, is a consequence of the difference which exists between the system on which coast lights are maintained in the United States and the system on which such lights are maintained in the United Kingdom.

In the United States the coast lights are erected and maintained by the federal government, and the expenses connected with those lights form part of the general expenditure of that government. The federal government, therefore, has a right to determine whether it shall reimburse itself for this outlay by levying light dues upon shipping, or whether, on the whole, it may not be more for the advantage of the United States, and more conducive to the commercial prosperity of the nation, that this charge should be borne by the public revenue, and that the commerce of the Union should be freed from any burthen in the shape of light dues upon vessels frequenting the ports of the Union. The government of the United States having power to decide this question, has determined, wisely, as her Majesty's government think, as well as liberally, to free the commerce of the Union from any burthen on this account, and to defray out of the national revenue the actual cost of erecting and maintaining the coast lights.

If the coast lights of the United Kingdom had been established upon the same principle, and if they had been erected and were maintained at the public expense, and if they were managed and administered by a department of the State, it is possible that her Majesty's government might think that it would be best for the general interests of the nation, that the system of the United States in regard to these matters should be adopted in this country, and that the commerce of the United Kingdom should be relieved from the burthen of light dues.

But the British government has not the power to deal with this matter as it pleases. The various lights which are established round the coasts of the United Kingdom have been erected and are maintained by various corporate bodies ; and those corporate bodies are entitled, by patents and by acts of Parliament, to levy certain dues upon shipping, in order to raise the necessary income for paying interest on the capital laid out in the construction of the lights, and for providing the means requisite for defraying the expense of maintaining those lights.

Her Majesty's government have no right or power to order these corporate bodies to abstain from levying these dues ; and these dues could not be made to cease unless Parliament was to vote such sums as would be necessary to buy up for the public the interest which the private parties concerned have in these lights ; nor unless Parliament were at the same time to authorize the government to abolish light dues for the future, and were to charge upon the public revenue the expense of maintaining the lights.

The expediency of adopting such a course has, indeed, from time to time, been suggested, but the question has not hitherto been considered with a view to any practical decision.

Her Majesty's government, however, cannot admit that the differ ence which exists between the system which prevails in the United States and that which subsists in Great Britain, in regard to coast lights, has the effect of infringing upon that principle of commercial reciprocity between Great Britain and the United States which forms the basis of the treaty of 1815. It is no part of the engagements of that treaty, that the internal system and local arrangements of the two countries upon commercial matters shall be the same. But the principle distinctly laid down in the second paragraph of the 1st article of the treaty of 1815, is, that the vessels of each country shall, in the ports of the other, be treated in regard to duties and charges, in the same manner and on the same footing as national vessels ; and this stipulation is strictly observed in regard to the light dues which are levied upon American vessels in British ports, for no other or higher light dues are levied in those ports upon American vessels than are levied in those ports upon vessels belonging to the United Kingdom.

But if the British light dues cannot be deemed to be any infringement of the principle of reciprocity which forms the basis of the treaty of 1815, neither can they be considered as in any degree conflicting with the liberal principle upon which the present navigation law of Great Britain is founded. For that navigation law relates to the admission of foreign ships into British ports with certain goods on board, and coming from certain voyages, which goods and voyages would before the passing of that law, have involved an exclusion from a British port ; but that law has no reference to light dues or harbor dues, nor does it make any mention of such matters, and it cannot, therefore, be appealed to as requiring for its complete execution, that any change should be made in such matters.

Her Majesty's government are quite ready to discuss with the government of the United States any question which may arise in regard to any supposed incompleteness in the mutual application of that principle of reciprocity in matters of navigation which is contained in the act 12 and 13 Vict., cap. 29 ; but willing and desirous as they are to carry out the provisions of that act to the fullest extent with respect to all countries which are disposed, as the United States have declared themselves to be, to pursue a similar course, yet her Majesty's government cannot admit the force of arguments founded upon a constructive application of that law to matters which are wholly beyond the range of its enactments.

The undersigned, &c.   (Signed)   PALMERSTON.

No. 3.

Mr. Lawrence to Viscount Palmerston.

UNITED STATES LEGATION,
138 PICCADILLY, *February* 12, 1851.

The undersigned, Envoy Extraordinary and Minister Plenipotentiary of the United States of America, has the honor to acknowledge the receipt of the note of Viscount Palmerston, Her Britannic Majesty's Principal Secretary of State for Foreign Affairs, of the 6th instant, in reply to the former note of the undersigned to Lord Palmerston, on the subject of the imposition of light dues on the tonnage of the United States within the ports of the United Kingdom, and will not fail to transmit a copy to his government by the earliest opportunity.

The undersigned avails himself of this occasion to correct an error into which Her Majesty's government seem to have fallen with reference to the request made by the undersigned, on behalf of the government of the United States, on this subject.

The government of the United States is not unaware of the difference between the system on which coast lights are maintained in the United States and the system on which such lights are maintained in the United Kingdom. On the contrary, the undersigned alluded to that difference in the former communication he had the honor to make to Viscount Palmerston with reference to them. That fact does not, however, diminish the pressure of this tax upon the shipping of the United States. And while the undersigned disclaims all intention of discussing the particular mode by which the lights of the United Kingdom are maintained, he still cherishes the hope that there is nothing in that system to prevent the changes for which his government have asked.

The federal government does not rest this request on the provisions of the convention of 1815. Lord Palmerston justly says, that "it is no part of the engagements of that treaty, that the internal system and local arrangements of the two countries upon commercial matters shall be the same."

Neither does it seek to view the present navigation law of the United Kingdom as liberating the commerce of the United States from this tax. Were that the case, there would have been no occasion to instruct the undersigned to make the communication of the 31st of December last.

The federal government rests this request solely on the fact that the commerce of Great Britain enjoys without charge the lights, beacons, and buoys maintained by the United States on a coast several thousand miles in extent; while the commerce of the United States is heavily taxed for the support of similar lights, beacons, and buoys in the United Kingdom. In view of this fact, it asks Her Majesty's government to meet the liberality of the United States with a reciprocal liberality. And as an additional reason for granting the request, it points to the present commercial policy of the two nations, founded professedly on the principle of reciprocity; and it invites Her Majesty's government to extend the principle still further, and treat the commerce of the United States as liberally as the United States treat the commerce of the United Kingdom.

The undersigned trusts that no question will arise in regard to any supposed incompleteness in the mutual application of that principle of reciprocity in matters of navigation contained in the act 12 and 13 Vict., cap. 29, to render necessary the discussion for which Viscount Palmerston expresses the readiness of Her Majesty's government.

The undersigned has great pleasure in learning that the expediency of adopting such a course as that of the United States has been from time to time suggested, and renews the expression of the hope that Her Majesty's government may devise some way to remove or to lighten this burden, which now falls so heavily on the commerce of the United States.

The undersigned, &c.,

ABBOTT LAWRENCE.

No. 4.

Viscount Palmerston to Mr. Lawrence.

FOREIGN OFFICE, *February* 14, 1851.

The undersigned, &c., has the honor to inform Mr. Lawrence, &c., that he has received and has referred to the proper department of Her Majesty's government, the note which Mr. Lawrence addressed to the undersigned on the 12th instant, relative to the light dues which are levied on American shipping in the ports of the United Kingdom.

The undersigned, &c.,

PALMERSTON.

REPORT OF THE LIGHT-HOUSE BOARD

TO THE SECRETARY OF THE TREASURY,

In answer to a resolution of the Senate of February 1, 1858, calling upon the Department for information in regard to the expense of erecting light-houses, &c.

TREASURY DEPARTMENT,
Office Light-house Board, March 13, 1858.

SIR : I have the honor, by direction of this board, and in compliance with the directions of the department, respectfully to submit the following report and accompanying papers, in answer to the resolution of the Senate of the United States, calling upon the Secretary of the Treasury to communicate to the Senate the annual expense of erecting light-houses and supporting the light-house system since the creation of the Light-house Board; and also the expense of the same number of years preceding the organization of the said board.

The period embraced in the clause of the resolution calling for the expense of erecting light-houses and supporting the light-house system, prior to the organization of the Light-house Board, is 5¼ years, viz: from the commencement of the fiscal year on July 1st, 1847, to the 30th September, 1852, inclusive; and the same period of time since the organization of the Light-house Board is from October 1, 1852, to December 31, 1857, inclusive.

The table hereto appended, marked A, exhibits the number of light-houses and lighted beacons; number of light-vessels and lights on board of them; expenditures under the several heads for each year and fraction of a year; the mean average rates of cost per annum of the lights, and the mean annual expenditures on account of the buoy service, and the amount of commissions paid to collectors of customs acting as superintendents of lights, for the 5¼ years immediately preceding the organization of the Light-house Board.

Table B is an exhibition similar to table A, for the period of 5¼ years under the management of the Light-house Board.

Table C exhibits the annual and aggregate special appropriations for new aids to navigation on the Atlantic, Gulf, and Lake coasts, and restoring old ones, for the period embraced in the resolutions of the Senate, immediately preceding the organization of the Light-house Board.

Table D is the same as table C, excepting that it embraces also the Pacific coast, and is for the period embraced in the resolution of the Senate since the organization of the Light-house Board.

Table E exhibits the amounts and balances of special appropriations on account of new aids and renovating old ones, authorized by Congress, available for those purposes on the 1st January, 1858, and the amounts which have reverted to the surplus fund.

Table F exhibits the expenditures for the support and maintenance of light-houses and buoys on the Pacific coast of the United States, to the 31st December, 1857, under the direction of the Light-house Board.

Table G exhibits the amount of balances in the treasury and available on account of the appropriations for the support and maintenance of the light-house establishment, at the close of the fiscal year ending June 30, 1857, and a similar list of balances to the 31st of December, 1857.

Table H exhibits a recapitulation of tables A and B, showing means of expenditures per annum and per light for the two periods of time preceding and succeeding the organization of the Light-house Board.

To which is appended "List of light-houses, beacons, and floating lights of the United States in operation on the 1st July, 1851, &c., carefully revised and corrected, by order of Stephen Pleasanton, Fifth Auditor and general superintendent of lights," (marked I,) and "List of light-houses, lighted beacons and floating lights of the United States, prepared by order of the Light-house Board, corrected to January 1, 1858," (marked J.)

From the tabulated statements embraced in these tables, it will be seen—

1. That the mean annual average cost of each light-house and lighted beacon, for the $5\frac{1}{4}$ years immediately preceding the organization of the Light-house Board, the mean average cost of oil being, for the same period, $1 13$\frac{3}{100}$ per gallon, was $1,302.

2. That the mean average annual cost of each light-house and lighted beacon for the $5\frac{1}{4}$ years under the management of the Light-house Board, the mean annual cost of oil for that period being $1 62$\frac{11}{100}$ per gallon, was $1,286.

3. That the annual average cost per light-house and lighted beacon, under the administration of the Light-house Board, has been $16 less than under the previous management for the same period of time; the difference in the average cost of the oil for illumination at the same time being $0 49$\frac{8}{100}$ per gallon greater since the organization of the Light-house Board than for the same period immediately preceding the organization of the Board.

The 325 light-houses and lighted beacons, existing at the date of the organization of the Light-house Board could not have been classed (according to established denominations, taking their power and range into consideration, in comparison with lights elsewhere) higher than—

   1 First class, or primary seacoast light.
   2 Second class, or secondary seacoast lights.
  16 Third class, or bay, sound, lake coast, &c., lights.
  87 Fourth class, or bay, sound, river, and harbor lights.
219 Fifth and sixth class, or river, harbor, and pier-head lights.

———
325

Of that number (325) there were—

One 1st order catadioptric or Fresnel apparatus.

Two 2d   "   "   "   "

One 3d   "   "   "   "

One 4th "   "   "   "

The others (320) were fitted with inferior reflectors and lamps, consuming, according to the estimates submitted to Congress for the fiscal year ending June 30, 1852, (page 65—A,) 106,365 gallons of oil per annum, as per statement, viz: Estimates for oil, &c., for fiscal year ending 30th June, 1852.—(Estimates, page 65—A.)

"For 331 light-houses, 3,093 lamps, 35 gallons each, 108,225 galls."
From which deduct for 6 reflector lights, difference
    between 331 and 325, at an average of 9 lamps
    each, 54 lamps, at 35 gallons each···· ········   1,890   "

Making total quantity for 325 lights············· 106,365   "
as found by the Light-house Board, according to the estimates submitted to the Department and to Congress.

Of the 320 reflector lights existing at the time of the organization of the Light-house Board but six remain to be fitted, or the apparatus provided for them, on the catadioptric system, which apparatus do not deteriorate from use nor require to be renewed, producing, according to the experience of all countries into which they have been introduced, at least four times as much light for the benefit of the navigator as the best system of reflector lights which has been devised, and, at the same time, at a consumption of not more than one-fourth of the quantity of oil required for the best system of reflector lights.

In illustration of the comparative merits and advantages of the two systems of light-house illumination, (reflectors in use prior to the organization of the Light-house Board and the catadioptric or lens system nearly completed under the management of the Light-house Board,) the following remark from a recent publication of British parliamentary papers "On the comparative merits of the catoptric and dioptric lights for light-houses," may be cited:

"The illuminating power of the most perfect kind of lenticular apparatus of the first order and the most perfect kind of parabolic reflectors are in the ratio of at least eight to one."

In further illustration of this subject, the estimate for oil for 331 lights, submitted to Congress for the fiscal year ending June 30, 1852, was 108,255 gallons, (Annual Estimates, page 65—A,) and the estimates for the fiscal year ending June 30, 1853, for oil for 349 lights, · was 114,520 gallons, (Annual Estimates, page 67—A,) which was at least one-seventh less than the actual quantity required for keeping efficient lights, with lamps and reflectors, as may be seen by referring to the excess of expenditures over appropriations, (table A, for the five and a quarter years anterior to the organization of the Lighthouse Board,) and from the fact that large quantities of oil were purchased and delivered to the different keepers by the superintendents, compared with the estimate for oil, (Estimates for 1858–'59, page 96—A,) "for 556 light-houses and lighted beacons, 48,150" gallons, under the management of the Light-house Board.

During the last four and a quarter years the sum of $155,479 07 has been expended by the Light-house Board from the appropriations for renovations, repairs, &c., of light-houses, for the purchase of the catadioptric apparatus referred to, for the lights existing at the time the board took charge, which was rendered indispensable in executing the law of Congress, of March 3, 1851, and to render the lights efficient, reliable, and economical. A deduction of this sum from the gross expenses for support and maintenance would reduce the average annual cost per light-house and lighted beacon under the management of the Light-house Board, from $1,286 to $1,195, or a difference in favor of the Light-house Board's management over that of the five and a quarter years previous to its organization of $107 per annum per light, and this, too, during a period of time when the most important item of light-house consumption cost one-third more than during the previous period of time with which the comparison is made.

The cost of other supplies, materials, and labor of all kinds, reached, during the last five and a quarter years, an equally great advance over the previous period, but which has not been taken into the account.

Another element of legitimate deduction in the expense of maintenance of the light-houses, under the Light-house Board, but which has not been taken into account, is the excess of expenditures of the first quarter of the fiscal year 1852–'53, immediately preceding the organization of the Light-house Board, in proportion to the gross sum appropriated for the entire year, (table A, column one-quarter year, 1852,) is the deficiency of supplies for the then current year, rendering the purchase of 21,000 gallons of oil, at a cost of $26,000, and other supplies for the lights indispensable. Comparing this deficiency with the supplies on hand available for the service during the next fiscal year, 1858–'59, under the Light-house Board, we find that there were in store, and available for the service of the next fiscal year, at the close of the deliveries for the current year, 35,000 gallons of oil, and other necessary supplies in like proportion, which, if deducted from the gross amount of money actually expended, would greatly reduce the average annual cost.

During the existence of the Light-house Board, fog bells and other fog signals have been authorized by Congress, including those previously authorized but not erected, amounting to $58,900. The placing of each of these bells or fog signals involved an expense of an additional light keeper to work it, or an increase of the salary of the keeper of the light-house at which placed, for the additional responsibility and labor incurred.

Lest it might be inferred that the condition of the towers and buildings, and the reliability and powers of the different lights, at the time of the organization of the Light-house Board and at the present time, were the same, it is deemed proper to recur to the number and classes, or order, of lights then and now.

	1st order.	2d order.	3d order.	4th order	5th and 6th order.	Total.
Prior to Light-house Board,	1	2	16	87	219	325
Under Light-house Board, December 31, 1857····	26	21	40	173	199	*459

4. In table A, under the head of light-vessels, the mean annual average cost per light for the $5\frac{1}{4}$ years prior to the organization of the Light-house Board is shown to have been $2,749.

In table B, under the head of light-vessels, the mean average annual cost per light, for the $5\frac{1}{4}$ years under the management of the Light-house Board, is shown to be $2,796. The mean average cost of oil purchased in the first named period (table A) being $1 $13\frac{3}{100}$ per gallon, and in the latter, (table B,) under the Light-house Board, being $1 $62\frac{11}{100}$ per gallon, making an excess of expenditure per

° Of this number six require lens apparatus to be provided.

17

light-vessel light per annum, under the management of the Light-house Board, of $47.

The aggregate amount of expenditures for support and maintenance of the light-vessels, from which the average annual cost per light is found, includes the building of four new light-vessels, to take places of old ones, and of 25 lanterns and reflector apparatus of the most approved description, for the light-vessels stationed at prominent points requiring the best lights that can be produced from light-vessels, to render the navigation of the localities safe and easy, and which expenditures were in addition to the ordinary repairs, refitments, &c., which amount in the aggregate to not less than $100,000.

Of the 34 light-vessels, containing 44 lights, existing at the time the Light-house Board took charge, there was but one of the 1st class, in tonnage or power of light, occupying a primary or exposed position; six of the 2d class, and the remainder, (27,) occupying unexposed positions, of small tonnage, and requiring small crews to take charge of them.

Of the 52 light-vessels, containing 72 lights, existing December 31, 1857, under the management of the Light-house Board, there were 11 of the 1st class, of 240 to 275 tons each, occupying exposed sea positions, requiring expensive outfits of anchors, cables, &c., and crews of about three times the number required by light-vessels occupying unexposed positions in bays, sounds, &c.; 12 of the 2d class, and the remainder occupying unexposed positions in bays, sounds, and rivers.

Within the last five years the wages of seamen in the navy has been increased from $12 to $18 per month, while the rates in the mercantile marine, to which the light-vessel service had mainly to look for crews, ranged at still higher figures. Rations which cost in 1852, and prior to that time, for the crews of light-vessels, from 19 to 20 cents per man per day, have averaged, during the last five years, from 25 to 35 cents per day per man. Labor and materials of all kinds for repairing light-vessels, and supplies other than oil, have advanced in proportion to the price paid for that article.

5. The mean annual average cost of the buoy and beacon service, (table A,) for the 5¼ years immediately anterior to the organization of the Light-house Board, was $75,664 60, and for a similar period of time, under the Light-house Board, it was $82,267 13. (Table B.)

The greater economy in this branch of the light-house establishment service, under the management of the Light-house Board, will be seen by referring to the fact that, prior to the organization of the board, the 6th section of the act making appropriations for light-houses, &c., approved September 28, 1850, which directs that all the buoys "shall be colored and numbered" as therein prescribed, was entirely neglected and disregarded ; and that in the general appro-priation bill for the support and maintenance of lights, &c., approved August 31, 1852, the first appropriation of $12,000 was made to carry out that act according to its terms.

The condition of the beacon and buoy service at the time of the organization of the Light-house Board as compared with its present state, the large increase in the number and improvements in the character of those aids to navigation, authorized by Congress to be placed since the organization of the Light-house Board, (table D, column special appropriations for buoys and beacons, amounting to $448,386 60 during the last 5¼ years,) and disregarding the large amount of property on hand available for this branch of the light-house service, and which is indispensably necessary for its economical and efficient management, the comparison will be found to be very favorable to the last 5¼ years.

6. In the column of table A, for the mean annual average amount paid to collectors of customs acting as superintendents of lights for the 5¼ years anterior to the organization of the Light-house Board, will be found $9,882 11, and the aggregate amount for the same period, under the same management, (i. e., prior to the Light-house Board,) $52,358 61.

In table B, under the same heading, the mean annual amount paid was $5,529 52, and the aggregate amount paid under the manage-ment of the Light-house Board was $28,847 66, making an annual saving, under the Light-house Board, of $4,352 59, and an aggregate saving for the 5¼ years of $23,510 95.

7. Table F exhibits the expenditures under the different heads of appropriation for the light-house service on the Pacific coast. The appropriations for that coast have been made upon estimates distinct from those for the Atlantic, Gulf, and Lake coasts, and as there were no aids on that part of the coast of the United States existing at the time of the organization of the Light-house Board, there were no prior expenses to be compared with them. The great distance from the Atlantic to that coast, and the difficulties and expenses attending the distribution of supplies to the lights there, render it necessary to keep a larger proportional supply of oil, &c., in store for future use

than on the Atlantic side. The cost of labor, materials, and supplies of all kinds has been, and is yet, three to five times what it is on the Atlantic coast, while the average rate of compensation of light-house keepers has been fixed by Congress at double the rate on the Atlantic coast.

8. Table C exhibits the amounts of appropriations, under the respective heads, for new aids to navigation, and for renewing old ones, made by Congress in special bills, from March 3, 1847, to August 31, 1852, and prior to the organization of the Light-house Board, amounting in the aggregate to $2,541,862 66.

Of those appropriations a number of the works remained to be completed, commenced, or condemned under the law as unnecessary, by the Light-house Board at the time it was organized.

9. Table D exhibits the amounts of appropriations, under the respective heads, for new aids to navigation and for renewing old ones, specially authorized by Congress, from March 3, 1853, to March 3, 1857, and during the existence of the Light-house Board, amounting to $3,636,930 72. Of these sums the appropriations made respectively on the 3d of March, 1853, 1855, 1857, amounting in the aggregate to $922,467 03, were based upon estimates in the annual estimates submitted by this board, and included by the Secretary of the Treasury in the annual estimates submitted by him to Congress. Those for the years 1854 and 1856, amounting in the aggregate to $2,714,463 69, were embraced in special light-house appropriation bills, originating with the Committees on Commerce of Congress.

10. Table E shows the sum of $1,756,205 81 unexpended, including $369,597 90 carried or to be carried to the surplus fund of the treasury, and $1,356,200 63 available on account of special light-house works authorized by Congress.

11. Table G shows at the close of the last fiscal year a total balance in the treasury of $467,015 49, exclusive of sums in the hands of disbursing officers available for the support and maintenance of the light-house service during the current year, and being that amount less than the sum appropriated or available for the general service, and also a balance at the close of the half of the current fiscal year (December 31, 1857) of $967,106 15 available for the remaining half and for the next year's service in maintaining the light-house establishment.

12. The table H is a recapitulation of the averages for the two periods of five and a quarter years each, both before and since the

organization of the Light-house Board, prepared from the tables before recited.

13. The two light-house lists, July 1, 1851, and December 31, 1857, will afford a general comparative view of the service at the two periods of time, and the columns of "built," "rebuilt," "refitted," of the latter will show in brief what has been done towards rendering the lights efficient and reliable by the Light-house Board.

It may not be amiss to add that the light-houses, lighted beacons, and light-vessels, authorized prior (but not built) and those authorized since the organization of the Light-house Board, amount in the aggregate to near 300; permanent beacons about 80; and the buoys have been increased within the same period nearly or quite four-fold.

The Light-house Board, in submitting its estimates, for the first time, (November 10, 1852,) for the support of the light-house establishment for the fiscal year ending June 30, 1854, states, in the letter accompanying them: "The estimates of this board for the fiscal year ending June 30, 1854, are the same in every respect as those for 1852, 1853, for the same objects. The additional estimates for objects authorized by the acts of March 3, 1851, and August 31, 1852, not contained in former lists and estimates, are based upon the same data, and bear relatively the same proportion to them.

"The additional estimates submitted for objects deemed of importance are not such as have hitherto been classed under the ordinary heads of repairs, &c., and amount in the aggregate to $27,000 less than the estimates for similar objects last year.

"For support of the light-houses and other aids to navigation on the coasts of California and Oregon, estimates are now submitted for the first time.

"The continued high prices of labor, &c., on the Pacific coast rendered it necessary that a different scale of estimating should be adopted for that coast; but in doing so the board has conformed its estimates to the most economical rates which would seem to be justified by the best information that could be obtained."

The letter of the Light-house Board of October 7, 1857, addressed to the Secretary of the Treasury, submitting estimates for the support of the light-house establishment for the fiscal year ending June 30, 1859, states:

"These estimates have been prepared to meet the actual state of the light-house service as it will be at the close of the present fiscal year, and not upon the pro-rata of expenditures of previous years, as heretofore, in view of the fact that by the commencement of the next fiscal year the system of catadioptric illumination authorized by the 7th section of the act of Congress making appropriations for light-houses, &c., approved March 3, 1851, and which has been in steady progress of execution since the organization of this board on the 9th October, 1852, will be near its full completion, which will thenceforth produce the economical results indicated at that time by greatly diminishing the annual consumption of oil, wicks, chimneys, and other supplies, as compared with that of the old system of reflectors and lamps, in addition to other benefits arising from increased brilliancy and power of the lights and from illuminating apparatus which is not liable to any sensible deterioration from use.

The aggregate amount of estimates submitted for the fiscal year ending June 30, 1859, for the Atlantic, Gulf, and Lake coasts, is ························· $712,598 99

The aggregate amount of estimates for the Pacific coast for the fiscal year ending June 30, 1859 ··········· 78,535 91

The aggregate amount of estimates for the fiscal year ending June 30, 1859, for the Atlantic, Gulf, Lake, and Pacific coasts, is ························· $791,134 90

Showing a diminution of ························· $399,471 39 in the estimates for the fiscal year ending June 30, 1859.''

The estimates for annual expenditures for support and maintenance of the light-house establishment, under the management of the Light-house Board, for the five fiscal years ending June 30, 1858, have been made at the same rate as that for the fiscal year ending June 30, 1853. The letters accompanying the estimates from year to year show this. In every case the existing light-houses, and those authorized to be built, were included. The object of this was to complete the renovation of the light-houses, and their equipment with Fresnel lenses, as soon as practicable, without asking Congress for special appropriations for the purpose.

By the end of the present fiscal year that object will have been accomplished, and it will be seen, from the letter of the board, of October 7, 1857, previously quoted, that the estimates for the fiscal year ending June 30, 1859, are based upon the saving made by the introduction of the lens system, and are the first fruits of that system, so far as regards an annual diminution of the expense of the establishment, the benefits of the introduction having been felt in all other respects since its commencement. A further diminution in the estimates may confidently be expected for the fiscal year ending June 30, 1860, when it is hoped that the expenditures will be brought to the minimum.

Notwithstanding the fact that large expenditures for rebuilding light-houses and purchasing new illuminating apparatus have been made from the general fund for support and maintenance, it will appear, by a comparison of the two periods of $5\frac{1}{4}$ years before and after the organization of the Light-house Board, that in the former period the expenditures overran the appropriations by $127,421 79, (a deficiency made good by transfers from special appropriation for light-houses,) while in the latter the appropriations exceeded the expenditures by $590,176 18.

Inasmuch as the subjects relating to light-houses, illumination, the management of the light-house service of this and other maritime countries, &c., were much discussed in Congress, from about 1838 to the passage of the law authorizing the organization of the Light-house Board in 1852, for a general view of the condition of the light-house establishment prior to the latter date the board would respectfully refer to the following congressional documents, being a part only of those printed :

Senate document No. 138, 2d session, 25th Congress.
Senate document No. 258, 2d session, 25th Congress.
Senate document No. 159, 2d session, 25th Congress.
Senate document No. 506, 2d session, 25th Congress.
Senate document No. 474, 1st session, 26th Congress.
Senate document No. 619, 1st session, 26th Congress.
Senate document No. 488, 1st session, 29th Congress.

Senate executive document No. 28, 1st session, 32d Congress, pages 18 to 20, *et seq.*

Senate executive document No. 22, 2d session, 32d Congress, pages 70, *et seq.*

House document No. 24, 3d session, 25th Congress, page 2, (oil tests, &c.,) and pages 48, 69, 70, 71, *et seq.*, and 113.

House document No. 183, 2d session, 27th Congress.

House executive document No. 114, 1st session, 32d Congress, and also, for a general view of the condition of the light-house service, under the management of the Light-house Board, to the several reports on the finances, submitted by the Treasury Department to

Congress, for 1853–'54–'55–'56, and to the report No. 16 in the finance report of December, 1857, from page 229.

It is respectfully submitted that the foregoing report and accompanying tables show the following facts :

1. The whole system has been remodelled according to the tenor of the 7th section of the act of Congress of March 3, 1851, producing the effects contemplated by that act with regard to economy and efficiency.

2. The number of buoys, beacons, and other day marks, has been increased by direction of Congress at least four-fold.

3. The number of light stations, since the organization of the Light-house Board, has, under the authority of Congress, been nearly doubled.

4. For the imperfect lamps and lanterns previously employed new apparatus has been introduced, the most perfect in character which the science and skill of the present day are able to afford.

5. Not only has a large diminution of the amounts of oil and other supplies for lights been effected, but the extent to which the seacoast lights are visible over the surface of the water has been greatly increased, which increase was indispensable for the safety of navigation.

6. From the combined results of these changes, the efficiency of the system has been multiplied eight times, at a nominal aggregate annual increase, the expenditures per light having been actually less than they were before the organization of the board.

7. This efficiency may be still further increased with an annual reduction of the expenditures, since the cost of the introduction of the new apparatus was much greater than that which will be required to continue its use.

Very respectfully,

W. B. SHUBRICK,
Chairman of the Light-house Board.

Thornton A. Jenkins,
W. B. Franklin. } *Secretaries.*

Hon. Howell Cobb,
Secretary of the Treasury.

TABLE A.—(ATLANTIC, GULF, AND LAKE COASTS.)

Exhibiting the number of light-houses and lighted beacons; rate of average annual cost of each light for supplies, repairs, keepers' salaries, and incidental expenses; total amount expended per annum for supplies, repairs, keepers' salaries, and incidental expenses of the light-houses and lighted beacons; number of light-vessels; average cost of support and maintenance per annum per light on board of light-vessels; total amount per annum expend'd for support, maintenance, and repairs of light-vessels; total amount per annum expended for buoy and beacon service; total amount of commissions paid to collectors of customs acting as superintendents of lights, &c., upon disbursements made by them for support and maintenance of the aids to navigation; total amount expended under the foregoing heads per annum for the five and a quarter years immediately preceding the organization of the Light-house Board in October, 1852, embracing the period from July 1, 1847, to September 30, 1852, and the rates and average paid for oil during that period.

YEAR.	Light-houses and lighted beacons.			Light-vessels.					Buoys and beacons.	Amounts paid to superintendents for commissions on disbursements.	Total amount expended for the support and maintenance of the light-house establishment.	Average cost per gallon of the oil purchased for each year of the lowest bidder under public advertisement.
	Number of lights.	Rate of average annual cost per light.	Total amount expended for supplies, inspections, salaries, repairs, and commissions of superintend'ts.	Number of light-vessels.	Average annual cost for support and repairs of each light-vessel.	Number of light-vessel lights.	Average annual cost per light on board of light-vessel.	Total am't expended for repairs, support, &c., of light-vessels.	Total am't expended for buoy and beacon service.			
1847–'48	259	1,229 00	315,862 02	80	$8,050 00	83	$2,406 00	$91,511 85	$61,997 67	$11,569 03	$471,971 54	$1 07.18
1848–'49	267	1,135 00	316,816 87	81	2,675 00	89	2,126 00	82,907 12	49,542 51	11,452 48	443,066 00	1 04.96
1849–'50	257	1,187 00	324,858 51	85	3,193 00	44	2,589 00	111,745 40	54,889 68	11,962 00	492,457 54	1 11.89
1850–'51	310	1,190 00	369,912 81	85	4,404 00	44	3,508 00	154,160 80	110,888 26	7,918 84	688,401 87	1 16.68
1851–'52	317	1,818 00	416,138 87	85	3,884 00	44	8,050 00	194,205 45	61,974 97	7,890 87	611,614 29	1 19.87
1st quarter of 1852–'53.	325	1,756 00	136,290 57	85	3,949 00	44	3,239 00	84,561 96	30,302 64	2,191 85	201,091 47	1 19.87
Mean annual average for 5¼ years.......	269	1,802 00	861,587 54	83	8,456 00	49	$2,749 00	119,800 81	75,664 60	9,989 11	548,580 42	1 18.08
Total amount of expenditures for 6¼ years, from July 1, 1847, to September 80, 1852....	......	....	1,882,304 45	.....	....	....	....	909,098 08	862,079 69	52,858 61	2,865,499 21	....
Total amount appropriated for 5¼ years, from July 1, 1847, to September 80, 1852.....	......	....	1,629,478 63	.....	....	....	....	931,639 29	887,016 73	61,095.09	2,742,249 13	....

W. B. SHUBRICK, *Chairman.*

THORNTON A. JENKINS, } *Secretaries.*
W. B. FRANKLIN,

TREASURY DEPARTMENT, *Office of Light-house Board, March 18, 1853.*

TABLE B.—(ATLANTIC, GULF, AND LAKE COASTS.)

Exhibiting the number of light-houses and lighted beacons; rate of average annual cost of each light for supplies, repairs, keepers' salaries, and incidental expenses; total amount expended per annum for supplies, repairs, keepers' salaries, and incidental expenses of the light-houses and lighted beacons; number of light-vessels; number of lights on board of light-vessels; average cost of support and maintenance per annum per light on board of light-vessels; total amount expended per annum for support, maintenance, and repairs of light-vessels; total amount per annum expended for buoy and beacon service; total amount of commissions paid to collectors of customs acting as superintendents of lights, &c., upon disbursements made by them for the support and maintenance of the aids to navigation; total amount expended under the foregoing heads per annum for the five and a quarter years immediately succeeding the date of the organization of the Light-house Board in October, 1852, embracing the period from October 1, 1852, to December 31, 1857, and the rates and average paid for oil during that period.

YEAR.	Light-houses and lighted beacons. Number of lights.	Rate of average annual cost per light.	Total am't expended for supplies, repairs, salaries, and commissions of superintendents.	Light-vessels. Number of light-vessels.	Average annual cost for support and repairs of each light-vessel.	Number of light-vessel lights.	Average annual cost per light on board of light-vessels.	Total am't expended for repairs, &c., of light-vessels.	Buoys and beacons. Total am't expended for buoy and beacon service.	Amounts paid to superintendents for commissions on disbursements.	Total amount expended for the support and maintenance of the light-house establishment.	Average rate per gallon of the oil purchased for each year of the lowest bidder under public advertisement.
1852-'53, for three quarters of year	325	$755 00	$154,652 87	39	$2,664 68	4?	$2,316 12	$81,649 46	$32,369 47	$8,083 44	$298,045 80	$1 29.23
1853-'54	339	1,442 00	487,299 09	40	3,893 53	52	2,610 64	185,758 29	56,438 63	6,083 12	679,586 01	1 89.10
1854-'55	408	1,494 00	609,670 01	45	4,246 18	61	3,192 42	191,078 19	106,421 55	6,275 22	907,170 83	2 06.00
1855-'56	434	1,119 00	435,917 25	51	3,862 96	71	2,774 80	197,011 20	84,500 53	6,247 78	761,429 94	1 97.25
1856-'57	459	1,206 00	558,498 66	52	3,793 55	72	2,786 18	197,005 03	94,661 08	5,999 79	645,169 74	1 51.00
December, 1857, for one-half of year	459	1,698 00	859,698 20	52	4,805 14	72	3,169 24	111,933 70	55,174 88	8,318 81	557,010 25	1 51.00
Mean annual average for 5¼ years	404	1,286 00	526,912 92	46½	3,749 55	62½	2,795 57	153,923 84	82,567 19	5,529 52	772,241 75	1 62.11
Total amount of expenditures for 5¼ years, from Oct. 1, 1852, to Dec. 81, 1857			2,710,241 18					914,424 89	429,634 62	25,847 66	4,054,300 69	
Total amount of appropriat'ns for 5¼ years, from Oct. 1, 1852, to Dec. 81, 1857			3,191,727 05					957,271 36	463,669 51	41,508 65	4,624,476 87	

THORNTON A. JENKINS, } Secretaries.
W. B. FRANKLIN,

TREASURY DEPARTMENT, Office Light-house Board, March 13, 1859.

W. B. SHUBRICK, Chairman.

TABLE C.

(ATLANTIC, GULF, AND LAKE COASTS.)

Exhibiting the amounts appropriated by Congress in special appropriation bills, reported from the Committees on Commerce and in the general appropriation bills for light-houses at new localities, rebuilding old light-houses, light-vessels for new localities, and rebuilding light-vessels, occupying old stations which required rebuilding, &c., for the five years (1847–1852) immediately preceding the organization of the Light-house Board.

Date of approval of appropriation bills.	Amount appropriated for new light-houses and rebuilding old ones.	Amount appropriated for new light-vessels and rebuilding old ones.	Amount appropriated for fog bells, &c.	Total.
March 3, 1847..........	$521,250 00	$25,000 00	....	$546,250 00
August 12 and 14, 1848.	252,091 90	64,000 00	$750 00	316,841 90
March 3, 1849..........	191,441 37	35,407 00	750 00	227,598 37
September 28, 1850.....	422,590 00	8,000 00	5,500 00	436,090 00
March 3, 1851..........	314,432 39	42,500 00	250 00	357,182 39
August 31, 1852........	495,200 00	130,200 00	32,500 00	657,900 00
	2,197,005 66	305,107 00	39,750 00	2,541,862 66

W. B. SHUBRICK, *Chairman.*

THORNTON A. JENKINS, } *Secretaries.*
W. B. FRANKLIN,

TREASURY DEPARTMENT,
    Office Light-house Board, March 13, 1858.

TABLE D.

(ATLANTIC, GULF, LAKE, AND PACIFIC COAST.)

Exhibiting the amounts of special appropriations made by Congress for erecting light-houses at new localities, rebuilding old ones, building light-vessels for new localities, &c., and for buoys, beacons, and fog bells for new localities, and restoring those destroyed, for the five years (1852 to 1857) immediately succeeding the organization of the Light-house Board.

Date of approval appropriation of bills.	Am'nt appropriated for new light-houses and re-building old ones.	Am'nt appropriated for new light-vessels.	Am'nt appropriated for fog bells, &c., for new localities.	Am'nt appropriated for buoys and beacons for new localities.	Total.
March 3, 1853.	$276,250 00	$28,000 00	$6,000 00	$43,160 00	$353,410 00
August 3, 1854.	1,210,338 00	33,500 00	19,600 00	239,640 00	1,503,078 00
March 3, 1855.	245,000 00	....	....	....	245,000 00
Aug. 18, 1856.	1,054,514 15	42,597 54	800 00	113,474 00	1,211,385 69
March 3, 1857.	231,838 81	°40,105 62	....	°52,112 60	324,057 03
	3,017,940 96	144,203 16	26,400 00	448,386 60	3,636,930 72

° To repair damages and supply losses occasioned by ice caused by storm of January 19, 1857.

W. B. SHUBRICK, *Chairman.*

THORNTON A. JENKINS, } *Secretaries.*
W. B. FRANKLIN,

TREASURY DEPARTMENT,
   Office Light-house Board, March 13, 1858.

TABLE E.

Exhibiting the amounts of special appropriations which were available on January 1, 1858, and of those which have reverted, or will revert, to the surplus fund, under the administration of the Light-house Board.

Balance on account of light-houses.............................	$1,356,200 63
Balance on account of buoys and beacons........................	30,407 28
Amount carried to surplus fund.................................	369,597 90
Total...	1,756,205 81

W. B. SHUBRICK, *Chairman.*

THORNTON A. JENKINS, } *Secretaries.*
W. B. FRANKLIN,

TREASURY DEPARTMENT,
   Office Light-house Board, March 13, 1858.

TABLE F.

(PACIFIC COAST.)

Exhibiting the amounts expended for support and maintenance of light-houses and buoys on the Pacific coast of the United States, from the times of their first exhibition to January 1, 1858.

Year.	Total amount expended for supplies, &c., for light-houses.	Total amount expended for repairs, &c., of light-houses.	Total amount expended for salaries of keepers and assistants of light-houses.	Total amount expended for beacon and buoy service.	Total amount expended for commissions of superintendents.	Total.
1853–'54	$10,790 12	------	------	------	------	$10,790 00
1854–'55	1,769 49	$1,874 95	$3,781 50	$1,424 00	------	8,849 94
1855–'56	31,820 26	16,745 17	15,220 91	4,083 05	------	67,909 33
1856–'57	50,757 14	6,284 45	13,773 09	6,367 90	$101 18	77,283 76
December 31, 1857, half year	18,840 45	11,596 22	9,526 27	1,197 86	------	41,160 80
Total amounts expended to December 31, 1857	113,877 40	36,540 79	42,301 77	13,072 81	101 18	205,893 95
Total amounts appropriated to December 31, 1857	162,038 63	58,094 50	124,000 00	44,250 00	2,700 00	391,083 13

W. B. SHUBRICK, *Chairman.*

THORNTON A. JENKINS, } *Secretaries.*
W. B. FRANKLIN,

TREASURY DEPARTMENT, *Office Light-house Board, March 13, 1858.*

TABLE G.

Exhibiting the balances remaining in the treasury, under the respective heads of appropriations, for the support and maintenance of the light-house establishment, at the close of the fiscal year ending June 30, 1857, and also at the close of the first half of the current fiscal year ending December 31, 1857.

	LIGHT-HOUSES.	LIGHT-VESSELS.	BUOYS.	Appropriation for expenses of inspections.	Appropriation for commissions of superintendents.	Total.
	Appropriations for supplies, repairs, and salaries of keepers of light-houses.	Appropriations for salaries of keepers, seamen's wages, repairs, &c., of light-vessels.	Appropriations for raising, cleaning, &c., buoys.			
Balances remaining June 30, 1857....	$357,941 49	$46,372 13	$46,563 76	$2,290 96	$13,847 15	$467,015 49
Balances remaining December 31, 1857.	678,047 29	153,526 87	114,604 19	2,930 33	17,997 47	967,106 15

W. B. SHUBRICK, *Chairman.*

THORNTON A. JENKINS, } *Secretaries.*
W. B. FRANKLIN.

TREASURY DEPARTMENT, *Office Light-house Board, March 13, 1858.*

TABLE H.

(RECAPITULATION.)

Exhibiting the average number of light-houses and lighted beacons, the average total annual expense of the light-houses and lighted beacons, average number of light-vessels, average annual cost per light-vessel, the average number of lights on board of light-vessels, the annual average cost per light, the average total annual expense of the light-vessels, the annual average expense of buoys and beacons, the average annual amounts paid to superintendents of lights for commissions on disbursements, the average total amounts of the cost of support and maintenance of the light-house establishment on the Atlantic, Gulf, and Lake coasts, and the average price of oil for 5¼ years immediately preceding and succeeding the organization of the Light-house Board, October, 1852.

	LIGHT-HOUSES AND LIGHTED BEACONS.			LIGHT-VESSELS.					BUOYS, ETC.	COMMISSIONS.	TOTAL.	OIL.
	Average number of lights.	Average annual cost per light.	Average amount expended for supplies, repairs, salaries, inspections, and commissions of superintendents.	Average number of light-vessels.	Average annual cost for support and repairs of each light-vessel.	Average number of light-vessel lights.	Average annual cost per light on board of light-vessels.	Average amount expended for repairs, support, &c., of light-vessels.	Average amount expended for buoy and beacon service.	Average amounts paid to superintendents for commissions on disbursements.	Average total amount expended for the support and maintenance of the light-house establishment.	Average cost per gallon of the oil purchased for each year of the lowest bidder, under public advertisement.
For the 5¼ years preceding the organisation of the Light-house Board.............	259	$1,802 00	$861,527 84	30	$6,456 00	42	$2,749 00	$118,800 81	$75,664 60	$9,862 11	$548,520 42	$1 18 08-100
For the 5¼ years succeeding the organization of the Light-house Board.............	404	1,256 00	526,912 92	46¾	8,748 55	62¾	2,796 57	155,928 84	52,267 18	5,039 59	772,247 75	1 69 11-100

THORNTON A. JENKINS, } Secretaries.
W. B. FRANKLIN,

W. B. SHUBRICK, Chairman.

TREASURY DEPARTMENT, Office Light-house Board, March 13, 1858.

St. Chéron, (Seine & Oise,)
May 7, 1861.

Sir : I was about leaving Paris, some three weeks ago, to locate myself in the country, where I pass the best part of the year, when I received the Report on Finances of the United States of 1858, as well as the very obliging letter with which you transmitted the document. I was then suffering from an attack of pleurisy, and my convalescence was retarded by several accidents. This circumstance, joined to the many embarrassments inseparable from a change of residence, will be my excuse for not having earlier expressed my lively thanks. § So soon as I was sufficiently restored to health I perused the important documents which you sent me, being attracted particularly by the chapters in relation to the light-house service. The prodigious development of this service within so short a time, under the Light-house Board, has truly astonished me. My old experience, in fact, enables me the better to appreciate how much energy and activity were necessary to bring to this degree of perfection the light-house service of such a vast expanse of coast—as well on the Pacific as on the Atlantic—without mentioning the task of succeeding in establishing, against hostile prejudices, the adoption of a new system. Much is due to you, sir, and to your honorable colaborateurs, for having created in so short a time this magnificent and combined establishment, and you should congratulate yourself that, thanks to your activity, the *Union, wherein is strength,* and which I find now so fatally compromised by the blindest passions, has not been overthrown before the accomplishment of your philanthropic work. I hope, however, that reason will yet triumph over these retrograde ideas, and that Providence will listen to the prayers of all generous hearts by maintaining the most admirable political structure which has ever been erected by the genius of liberty.

Excuse, dear sir, this digression upon the seething volcano of your national politics, and be pleased to accept the renewed assurance of my high esteem, as well as of my sentiments of great devotion.

LEONOR FRESNEL.

To Com'r Thornton A. Jenkins, U. S. Navy,
Secretary of the Light-house Board,
Washington city, U. S. A.

FINIS.

www.ingramcontent.com/pod-product-compliance
Lightning Source LLC
Chambersburg PA
CBHW021515210326
41599CB00012B/1257